CONGENITAL HEART DISEASE AND NEURODEVELOPMENT

T0282348

CONGENITAL HEART DISEASE AND NEURODEVELOPMENT

UNDERSTANDING AND IMPROVING OUTCOMES

Edited by

CHRISTOPHER McCUSKER

FRANK CASEY

ELSEVIER

AMSTERDAM • BOSTON • HEIDELBERG • LONDON
NEW YORK • OXFORD • PARIS • SAN DIEGO
SAN FRANCISCO • SINGAPORE • SYDNEY • TOKYO

Academic Press is an imprint of Elsevier

Academic Press is an imprint of Elsevier
125 London Wall, London EC2Y 5AS, UK
525 B Street, Suite 1800, San Diego, CA 92101-4495, USA
50 Hampshire Street, 5th Floor, Cambridge, MA 02139, USA
The Boulevard, Langford Lane, Kidlington, Oxford OX5 1GB, UK

British Library Cataloguing-in-Publication Data
A catalogue record for this book is available from the British Library

Library of Congress Cataloging-in-Publication Data
A catalog record for this book is available from the Library of Congress

ISBN: 978-0-12-801640-4

For information on all Academic Press publications
visit our website at https://www.elsevier.com/

Typeset by TNQ Books and Journals
www.tnq.co.in

Dedications

To Ursula for your constant support and love through all my work, and to Caroline and Tom for the joys and lessons of fatherhood that have made me a better psychologist.

—Christopher McCusker

To Pauline, and to my daughters Emma, Sinead, Aoife, and Claire for the love, support, and encouragement that makes everything possible.

—Frank Casey

Contents

I

HEARTS AND MINDS

1. Congenital Heart Disease: The Evolution of Diagnosis, Treatments, and Outcomes

F. CASEY

2. Historical Perspectives in Pediatric Psychology and Congenital Heart Disease

N. ROONEY

II

TOWARD A NEURODEVELOPMENTAL PHENOTYPE

3. A Longitudinal Study From Infancy to Adolescence of the Neurodevelopmental Phenotype Associated With d-Transposition of the Great Arteries

D.C. BELLINGER AND J.W. NEWBURGER

4. Neurodevelopmental Patterns in Congenital Heart Disease Across Childhood: Longitudinal Studies From Europe

H.H. HÖVELS-GÜRICH AND C. McCUSKER

5. An Emergent Phenotype: A Critical Review of Neurodevelopmental Outcomes for Complex Congenital Heart Disease Survivors During Infancy, Childhood, and Adolescence

M. KHARITONOVA AND B.S. MARINO

III

PSYCHOLOGICAL PROFILES AND PROCESSES

6. Is There a Behavioral Phenotype for Children With Congenital Heart Disease?

C. McCUSKER AND F. CASEY

7. A Family Affair

N. DOHERTY AND E. UTENS

8. The Adult With Congenital Heart Disease

D. KATZ, M. CHAPARRO AND A.H. KOVACS

IV

INTERVENTIONS

9. The Congenital Heart Disease Intervention Program (CHIP) and Interventions in Infancy

N. DOHERTY AND C. McCUSKER

Contributors

David Bellinger is a Professor of Neurology and Psychiatry at Harvard Medical School and a Professor of Environmental Health at the Harvard T.H. Chan School of Public Health.

Frank Casey (Editor) MD FRCP MRCPCH BSc is Consultant Paediatric Cardiologist at the Royal Belfast Hospital for Sick Children and an Honorary Senior Lecturer in Child Health at Queen's University, Belfast. He trained in Pediatric Cardiology in Belfast and at the Hospital for Sick Children, Toronto. He is the Clinical Lead for the newly developed All-Ireland Paediatric Cardiology Network. He is also the Chairperson of the Psychosocial Working Group of the Association for European Paediatric and Congenital Cardiology. For more than 20 years he has led a research program in Belfast focused on the biophysical, neurodevelopmental, and psychosocial outcome for children with congenital heart disease. This work has pioneered interventional programs to modify lifestyle behaviors in children treated for congenital heart disease including the landmark Congenital Heart disease Intervention Program described in this book. He has many publications from that research program and is recognized as an international leader in this area of work.

Maria Chaparro is a Clinical Psychology Intern, Toronto Congenital Cardiac Centre for Adults, Peter Munk Cardiac Centre, University Health Network in Toronto, Canada.

Nicola Doherty is a Consultant Lead Clinical Psychologist currently based in the Western Trust area of Northern Ireland where she leads up the Paediatric Psychology Service and the Clinical Psychology input to the Child and Adolescent Mental Health Service. Previously she led the Paediatric Psychology Service in Belfast Trust, based at the Royal Belfast Hospital for Sick Children (RBHSC). She holds an honorary lecturer post with Queens University, Belfast and is involved in various regional committees and working groups. Her first, post-qualification post was as research clinical psychologist for the Congenital Heart disease Intervention Program (CHIP) project based at RBHSC and has had ongoing clinical and research interest in pediatric cardiology, other pediatric presentations, neurodevelopmental conditions, feeding, and in infant mental health.

Hedwig Hövels-Gürich MD is a Professor and senior Pediatric Cardiologist and lecturer at RWTH Aachen University of Technology in Aachen, Germany. She works in the Department of Pediatric Cardiology at the Center for Congenital Heart Disease of the University Hospital. She has been a long-time active member of the *Psychosocial Working Group of the German Society of Pediatric Cardiology*. She is also a member of the *International Cardiac Collaborative on Neurodevelopment* (ICCON) *Investigators*. Since 1995, she has been at the European cutting edge of research, which has assessed the long-time outcomes of neurodevelopmental, psychosocial, and the quality of life aspects after corrective surgery for congenital heart defects in infancy.

Danielle Katz is a Clinical Psychology Practicum Student, Toronto Congenital Cardiac Centre for Adults, Peter Munk Cardiac Centre, University Health Network in Toronto, Canada.

Maria Kharitonova PhD is a Research Assistant Professor at the Medical Social Sciences department at the Northwestern University Feinberg School of Medicine, Chicago.

Adrienne Kovacs is a Clinical and Health Psychologist at the Toronto Congenital Cardiac Centre for Adults, Peter Munk Cardiac Centre, University Health Network in Toronto, Canada. She is also an Associate Professor in the Department of Psychiatry in the Faculty of Medicine at the University of Toronto and an Affiliate Scientist with the Toronto General Research Institute. As a clinician and funded researcher, she is committed to optimizing psychological outcomes and the quality of life in adults with heart disease.

Bradley Marino MD, MPP, MSCE is a Professor of Pediatrics and Medical Social Sciences at Northwestern University Feinberg School of Medicine, Chicago. He is Director, Center for Cardiovascular Innovation at the Stanley Manne Children's Research Institute and Co-Director, Neo-Heart Developmental Support Program. He is the Co-Director of Research and Academic Affairs within the Divisions of Cardiology and Critical Care Medicine at Ann and Robert H. Lurie Children's Hospital of Chicago.

Christopher McCusker (Editor) completed his Clinical Psychology training at the *Institute of Psychiatry, Kings College London* in 1991 after attaining a first class honors and PhD in Psychology from the *Queen's University of Belfast*. Since that time he has combined clinical practice with leading the Doctoral program for Clinical Psychology training in Northern Ireland and with research related to his clinical interests. He has been the chair of the professional body for Clinical Psychology in Northern Ireland and has led on developments in training standards for the profession within the United Kingdom. Since 2000 his research has focused on understanding the determinants of outcomes for children with chronic illness and their families. This has been across pediatric specialisms including acquired brain injury, epilepsy, neurodevelopmental disorders, and congenital heart disease. He has authored numerous publications in these domains and, most importantly, has translated knowledge of risk and protective factors into new programs of early psychological interventions to promote resilience in these children and their families. In this book he presents the

Congenital Heart disease Intervention Program (CHIP) which is the first of such programs internationally for these children and their families.

Margaret Louise Morrison BSc (Hons) MD MRCPCH is a specialist registrar in pediatric cardiology at the Royal Belfast Hospital for Sick Children. She has a special interest in promoting exercise capacity and quality of life in children and young people with congenital heart disease.

Jane Newburger is a Commonwealth Professor of Pediatrics at Harvard Medical School and the Associate Cardiologist-in-Chief for Academic Affairs in the Department of Cardiology at Boston Children's Hospital.

Nichola Rooney was appointed as the first dedicated clinical psychologist in the pediatric cardiology unit of the Royal Belfast Hospital for Sick Children, where she established and developed an internationally recognized clinical and research unit. Her research collaboration with the hospital's pediatric cardiologists, led by Dr. Connor Mulholland, attracted the initial funding for the establishment of the Congenital Heart disease Intervention Program (CHIP) described in this book. She is currently the Chair of the Children's Heartbeat Trust which in addition to providing support services to children and families, is working to establish an all-Ireland congenital heart disease service, including a multidisciplinary research development and support unit.

Elizabeth Utens is a clinical/child psychologist, Associate Professor at the Department of Child and Adolescent Psychiatry, and research coordinator Paediatric Psychology in the Erasmus MC-Sophia Children's Hospital Rotterdam. She has been undertaking and publishing research for more than 20 years into the psychosocial functioning and the quality of life in children, adolescents, and (young) adults with pediatric diseases such as congenital heart disease, meningococcal septic shock, cystic fibrosis, inflammatory bowel disease, children undergoing pediatric anesthesia, and children with psychiatric anxiety disorders, as well as their parents. She is a member of the steering committee of the Psychosocial Working Group of the European Association for Paediatric Cardiology, member of the International Society for Adult Congenital Heart Disease, and chair of the Dutch national network for Paediatric Psychology.

Preface and Overview

C. McCusker

The Queens University of Belfast, Northern Ireland;
The Royal Belfast Hospital for Sick Children, Belfast, Northern Ireland

F. Casey

The Royal Belfast Hospital for Sick Children, Belfast, Northern Ireland

THE IMPORTANCE OF STORIES

This is a book of narratives. Mostly it is a book about how pediatric psychologists and cardiologists across the world are constructing a narrative to help us make sense of the outcomes for children with congenital heart disease (CHD) and in a way which points toward intervention strategies to improve such outcomes. This narrative is rooted in the scientific tradition of empirical research. Many readers will be familiar with that tradition. However, we thought that we would start this book with the stories of the children that we have known and worked with—stories which we believe the various chapters of this book will help us better understand and, most importantly, help in such future scenarios.

Lily's Story—Christopher McCusker

I had a best friend at primary school called Lily. Sometime around primary three a routine school medical highlighted that all was not quite right with Lily's heart. So began a journey involving various medical investigations which would ultimately lead to open heart surgery to repair a "hole" in her heart.

I remember traveling from our small town up to the regional children's hospital in Belfast to visit Lily with her family. Although she seemed well enough, cheerfully showing me the various new toys and books she had accrued, something of a hush seemed to have descended around her. A nascent fear was emerging. Fear not to overdo it or put "pressure" on her "weak" heart would surround Lily for some years to come.

Something was lost. My best friend seemed different as she assumed the mantle of a delicate and special child. A minor cold led to days, even weeks, off school and multiple health consultations; physical education and running through the woods surrounding our home became imbued with danger. Imperceptibly, Lily retreated to the fringes of our group of friends. On the other hand, this youngest child took center stage at home. Lily's capacity for wrongdoing and need for discipline appeared to have evaporated. Mother, father, brothers, and sisters all now seemed to calibrate their lives around ensuring that Lily, this special child who might check out on them anytime, had the best life that they could offer.

Truth be told, Lily had an uncomplicated defect and later enjoyed what we would now recognize as a "good outcome." Her natural drive for autonomy, and shaking off the shackles of the "sick" child, broke through in the teenage years. Perhaps because of coming through what she did, of the lessons she learned, she has become a resilient, happy, and well-adjusted woman. Her congenital heart defect became a distant memory to most of her friends, although it continues to this day to be a part of her life routine, as she still attends the regional review clinic for adults with CHD. Now a grandmother, Lily has ultimately negotiated all the developmental transitions challenged by this disease well.

Katie's Story—Frank Casey

I first met Katie with her parents at the age of 2 days. No problem had been suspected antenatally but on the 2nd day of life she was found to be dusky and breathless. Katie was her parents' first child. Our tests at the first assessment told us that Katie had a complex heart problem with a single ventricle heart. Her parents were devastated as the details of the heart problem, the possible treatments, and the future prognosis were explained to them. In the 1st month of life Katie had her first operation, and in the following weeks she remained in hospital struggling with feeding and weight gain.

Despite all the struggles Katie's parents bonded with her and spoke openly about their sense of "grief" of what was happening to their child. From early on it was clear that Katie too was a strong character—she survived her second operation at the age of 11 months and then a third surgery at the age of 4 years. It was not possible to perform a "corrective" operation for her complex heart problem and so she had a "Fontan" operation.

Until the age of about 8 years Katie found she could keep up with her peers in physical activities but as she got older she was more conscious of the differences in physical capabilities. Being very keen on outdoor activities she was frustrated by this, leading to difficult behavior both at home and at school. In her early teenage years she further rebelled and stopped taking her medications, unknown to her parents. At that stage Katie had ongoing input from the Clinical Psychology Service over a period of a year and she also found a physical activity she loved in riding and caring for horses. She is now in her final year at school and hopes to go to University and become a teacher.

Last month her family organized the 18th birthday party for her to celebrate, as they put it, their gratitude for 18 years that they thought they might not have had. I was privileged to be at her party and heard her give a speech to those present, that acknowledged the challenges she and her family had faced and will have to face in the future. The key to Katie becoming a very accomplished young woman, after a turbulent journey, was her courage, her determination not to let her heart disease defeat her, the support of her parents who treated her as "normal," and the input of the multidisciplinary health-care team with appropriate interventions at the times they were most needed.

Other Stories of CHD

It is perhaps important to note from the outset that the outcomes for Lily and Katie are probably the norm for children with CHD. However, that journey is not always easy and the stories vary greatly for children and adults with CHD.

Medical advances have dramatically increased survival rates and improved physical outcomes for children with CHD. In the same time period there has been an emerging recognition within the pediatric cardiology community that true "success" of treatment needs to address all aspects of the child's well-being. In our practice we have been struck by the persistent observation that a significant proportion of these children struggle with their developmental transitions—at home, in friendships, and at school. Importantly, such outcomes have not appeared to be mostly related to the severity of the disease. Rather, we have been struck by the importance of families in determining such outcomes. This realization, borne out by much of the research to be discussed in this book, broadened our horizons about what, and how, to treat the children who passed through our unit.

At the *Royal Belfast Hospital for Sick Children*, Northern Ireland, the pediatric cardiology specialism was one of the first to recognize that psychological interventions should be an integral (and integrated) part of the service. In the early 1980s, the then lead pediatric cardiologist, Dr. Connor Mulholland, recruited the author of the second chapter of this book, Professor Nichola Rooney, to provide psychological consultation to the medical team and psychological assessments and interventions for the children and their families. So began a program of clinical and research activity which has led to much of the contemporary knowledge and insights contained in this book.

Through the years we have encountered and treated many children whose problems are illuminated by our research and that of the other experts from Europe and North America, contained in these chapters. Consider baby Molly, failing to thrive many months after the successful repair of her heart defect. Despite the best advice of the nurses and health visitor, Molly's mother could not get her to feed and herself had become despondent and depressed at her apparent incapacity to soothe and bond with her fractious and "jittery" baby. Or 6-year-old Malachy, who was not settling at all well into school. Clumsy and uncoordinated, he was not making friends easily and was struggling with preliminary literacy and numeracy targets. Consider the Jacobs family, chronic attenders at the emergency department and *out of hours* service, alarmed by perceived palpitations, breathlessness, and "poor color" of their 8-year-old son, Charlie, whose heart defect had been successfully repaired several years earlier. Despite such familiarity with health-care settings, this introverted and withdrawn boy would fly into a blind panic in the face of the most minor medical intervention.

Such problems, if left untreated or superficially palliated, would cumulatively build and increase the risk for problems in later childhood and adolescence. Grainne, an 11-year-old, whose parents' fear of allowing her to exercise had led to inappropriate restrictions over the years which had as a consequence led to reduced stamina and breathlessness. This was a particular issue in her school where the anxiety about exercise had been communicated to her teachers. This only served to reinforce a vicious cycle of inactivity and further health concerns.

Peter, a 13-year-old boy, seemed to be falling further and further behind at school, with each passing year. Although he had held his own in primary school, he seemed unable to manage the increasing demands for independent problem solving and organization required of him at high school. Mary, a 16-year-old girl, who, rather late in development, was being considered for a diagnosis of autistic spectrum disorder. Insidious features appeared to have crept up on the family, but promised to explain many years of difficulty in managing her behavior, poor peer relationships, and school underachievement.

OVERVIEW OF THE BOOK

These children reached some threshold for focused input from psychology services and most saw their problems understood and turned around. Our research and anecdotal clinical experience highlighted that countless more were passing through childhood just under the radar, but on the margins of "normal" development. It is now well established, however, that risk for psychopathology, poor adjustment, and underachievement cumulatively build. Thus problems at any one of the developmental stages, noted above, were likely to have had their seeds sown by perhaps less severe, but unresolved, earlier difficulties. In turn such problems, if left untreated, would amplify risk for later difficulties.

This book has been written by clinicians who have had daily experience of working with children with CHD and their families. Importantly, all have led programs of research which have helped us better identify the nature of the problems these children and their families face. Phenotypes, or identifiable clusters and patterns of developmental and behavioral features, have been emerging. These are discussed together with the factors we have learned amplify, or reduce, these difficulties. Authors will draw together the conclusions from their own work and that of others around the world in a way which improves understanding and ultimately care. The first early intervention program of its kind, aimed at preventing the seeding and escalation of problems, are outlined. Results have been both promising and exciting and offer some preventive strategies which can be woven into pediatric cardiology services around the world.

In chapter "Congenital Heart Disease: The Evolution of Diagnosis, Treatments, and Outcomes," Frank Casey provides an overview of the nature of CHD, how and when it is diagnosed, and the developments in medical and surgical treatments therein. He provides the reader with an essential understanding of the nature of congenital heart defects, which vary in terms of cyanosis (whether or not oxygenation of the blood is compromised), surgeries required (whether cardiopulmonary bypass is needed or not), and what conditions can be corrected as opposed to palliated only. These are factors that have been considered important in understanding outcomes for the child and the research outlined later in the book often references groups of children with CHD based on such taxonomies.

The next chapter by Nichola Rooney provides us with some historical context for current practice and research. Events leading to a "marriage" of the disciplines of pediatrics and psychology are described together with the foci for early psychological interventions with children with chronic illness and disability which arose from this. As more and more children with even complex CHD survived, Rooney describes the early research which examined outcomes in terms of quality of life, cognitive functioning, and behavioral adjustment. Findings, although often contradictory, generally highlighted distinct potential for negative sequelae, the need to look more closely but also the need for research to move away from assumptions of unidirectional influence (disease–outcome). Rather, more systemic models were called for, which recognize the interacting influences of disease, family, and environmental processes in determining outcome. Such models underpin the contemporary research discussed in this book.

In Part 2 we highlight that this is a disease which not only affects the heart but also the brain. Once anecdotal, the evidence has grown to suggest that there may in fact be a neurodevelopmental phenotype associated with CHD. David Bellinger and Jane Newburger profile the neuropsychological and psychological features of children with one particular but significant CHD, transposition of the great arteries, across the first 16 years of life. This seminal series of longitudinal studies highlights that while markers of neurodevelopmental dysfunction are present from infancy (most notably in motor domains), further cognitive problems emerge through childhood and adolescence. In particular, higher-order executive functioning deficits become apparent which they propose have associated consequences for pragmatic language, planning and organization and ultimately psychosocial adjustment. They present evidence from structural neuroimaging studies which implicate a biological basis to these difficulties.

In chapter "Neurodevelopmental Patterns in Congenital Heart Disease across Childhood: Longitudinal Studies from Europe," Hedwig Hövels-Gürich and Chris McCusker present work from Europe with children with a range of other significant congenital heart defects and surgical interventions. The story appears remarkably similar to the Boston trials with early motor deficits, which themselves sometimes recover, giving way to a wider range of problems with attention, memory, and integrative reasoning. The research suggests that these are significant enough to impact on academic and psychosocial outcomes for these children. Neurological frameworks for understanding recovery versus late effects, following insults to the developing brain, are considered. A four-factor model for considering the risk is discussed which involves concomitant neurological features, perioperative factors, preoperative hypoxemia, and family and environmental processes.

Finally in this part, Maria Karitonova and Bradley Marino systematically review the evidence for domain-specific cognitive deficits, from infancy through to adolescence, across all key studies which have contributed to this corpus of knowledge. This critical review confirms the preponderance of visuomotor problems across ages and the emergence of problems in later childhood with especially integrative aspects of attention, language, and memory. Executive functioning appears implicated as a common etiological deficit and an emergent phenotype appears evident. The authors present a case for the links between these neuropsychological features and academic and psychosocial difficulties. Routine neuropsychological screening is called for, but both clinical and research initiatives need to be better informed by analysis of the specific processes underpinning these outcomes.

Part 3 considers psychosocial outcomes for children and adults with CHD. In chapter "Is There a Behavioral Phenotype for Children with Congenital Heart Disease?" Chris McCusker and Frank Casey take a developmental psychopathology framework for understanding how and why difficulties may cumulatively build across childhood. An incisive critique of the equivocal evidence base to date is provided and reasons for "noise" in the research literature are considered. The authors present new evidence which suggests that the primary behavioral phenotype for these children relates to deficits in personal and interpersonal competencies, rather than with mood disturbance per se—although the latter may develop as secondary consequences. A multifactorial model of risk is presented, but the authors highlight the central role of maternal and family factors which, they argue, offers new routes for interventions to minimize maladjustment and promote better outcomes for the child with CHD and their family.

In the next chapter Nicola Doherty and Elizabeth Utens highlight that CHD is indeed a family affair. Their work highlights that family functioning, so important in determining outcomes for the child with CHD, is itself affected by CHD. However, they present research evidence which suggests that many of the key factors that mediate this, such as parenting skills, family processes, and collaboration in care, are things that we can actually do something about. Foci for interventions are suggested.

In the final chapter of this part Danielle Katz, Maria Chaparro, and Adrienne Kovacs remind us that most people living with CHD now are adults. Continuing the developmental challenge framework, the authors highlight the key developmental transitions of adulthood challenged by CHD. The evidence for psychological morbidity is equivocal, as in the childhood literature, with some showing better than average adjustment. However, on the whole, one in three appear at risk for adjustment difficulties. Again psychosocial rather than disease/medical factors appear to be most implicated in outcome. Perhaps more importantly, Katz and colleagues suggest we should focus less on pathology and more on the psychosocial challenges pertaining to identity, attaining independence, relationships and parenting, and reduced life expectations which, under certain contexts, potentiate the risk for maladjustment. Recommendations for interventions are considered.

In Part 4 we move to considering perhaps the most important question for clinicians and researchers—how might we translate this increased understanding of what determines outcomes for these children and their families into effective interventions? In chapter "The Congenital Heart Disease Intervention Program (CHIP) and Interventions in Infancy," Nicola Doherty and Chris McCusker introduce the *Congenital Heart disease Intervention Program* (CHIP). Given the importance of parents and families for determining outcomes for children with CHD, this first early intervention program for this population is not only aimed at empowering the family, but also targeted at key developmental transitions for the child and family. *CHIP–Infant* is described in this chapter, which is aimed at parents of infants with recent diagnoses. Interventions are aimed at bolstering parent–infant transactions through psycho-education, narrative therapy, problem-solving therapy, and specific coaching in strategies to improve neurodevelopment and feeding. Interventions were evaluated in a controlled trial with gains evident in the intervention group in terms of maternal mental health, family functioning, feeding behaviors, and infant neurodevelopment.

Chris McCusker in chapter "Growing Up: Interventions in Childhood and CHIP–School," demonstrates how the CHIP program can be adapted to different developmental stages, with different challenges for the child and family. *CHIP–School* uses the same principles of problem-solving therapy and meaning making but woven through these are specific interventions related to promoting independence, exercise and activity, and behavioral adjustment. Important outreach interventions to those community health and education services are also described as core components. Outcomes here were evaluated in a randomized controlled trial. Again interventions were associated with reduced psychological distress in mothers and improved family functioning. Positive benefits for the child were demonstrated including fewer days sick and off school, and a trend toward better behavioral adjustment.

Finally in this part, Louise Morrison and Frank Casey describe a different intervention program specifically targeted at improving exercise, health and well-being in teenagers with CHD. Again, psychological principles underpinned an intervention which included group motivational counseling interventions prior to formulating an individualized structured program of exercise training. Further positive findings were demonstrated in a randomized trial in terms of positive appraisals and confidence about exercise and actual increases in exercise capacity and daily exercise activity.

In the concluding chapter of the book Chris McCusker draws together the evidence presented and discusses the nature of the neurodevelopmental and psychological phenotype which is emerging. He further discusses those factors which we have demonstrated as important in mediating outcomes and proposes an integrative model of understanding which synthesizes these conclusions and offers a tool to aid clinical formulation and intervention. Important directions for future research are considered, together with the challenges posed for translating interventions into routine clinical practice across health-care systems. Finally, we revisit the stories of Lily, Katie, and the other children outlined above and consider what the clinical research outlined in this book tells us about their experiences.

Acknowledgments

We would like to thank Kristi Anderson and all the team at Elsevier for supporting this project and keeping us on track throughout many busy months when the demands of work threatened to derail us. We thank our authors from the United Kingdom, Europe, and North America who have shared their cutting-edge research and thinking. Your contributions will make this book a truly landmark publication and a beacon of good practice for pediatric cardiology services in the future. Finally, and most importantly, we would like to thank all the children with congenital heart disease, and their families, that we have worked with over the years. We have learned much from sharing your journeys—in dark days as well as good—and it is you who have inspired the work of this book.

PART I

HEARTS AND MINDS

CHAPTER

1

Congenital Heart Disease: The Evolution of Diagnosis, Treatments, and Outcomes

F. Casey

The Royal Belfast Hospital for Sick Children, Belfast, Northern Ireland

Congenital heart disease (CHD) is usually defined as a structural abnormality of the heart or intrathoracic vessels present at birth that is actually or potentially of functional significance.

CHD is the most commonly occurring congenital abnormality and affects about 8 per 1000 live births.[1] There is a wide spectrum of severity ranging from minor defects to very complex abnormalities that have a lifelong impact on the life of the child. Severe CHD is sometimes defined as CHD necessitating surgical intervention or causing death in the first year of life. Even in the modern era where most conditions are surgically treatable CHD still accounts for 5–10% of neonatal deaths.[2,3]

CLINICAL PRESENTATION OF CONGENITAL HEART DISEASE

The clinical presentation of CHD depends on the severity of the defect and its physiological effect.

FETAL DIAGNOSIS

In the current era many of those with major CHD may have the diagnosis made antenatally by fetal echocardiography (a detailed ultrasound examination of the fetal heart). In Great Britain and Ireland during 2013–2014, 46% of children born with CHD requiring surgery in the first year of life had the diagnosis made antenatally.[4] Diagnosis before birth gives the opportunity to counsel and prepare the family for the birth of the child with CHD and from a medical care viewpoint, allows for a planned delivery of the affected child in a pediatric cardiology center. Thus appropriate treatment can be instituted at the earliest possible point potentially reducing morbidity and mortality.

NEONATAL PRESENTATION

If not detected antenatally, those babies with severe CHD will usually present during the neonatal period. The newborn presentation depends on the type of defect and can be with cyanosis or with cardiac failure and shock. In the most severe conditions such as those with only a single ventricle the circulation may be dependent on the patency of the ductus arteriosus. This vessel normally closes within 24–36h of birth, and this classically is the time interval when such babies present. Others will be diagnosed after detection of a heart murmur on examination. Hoffman and Kaplan[5] found that 46% of cases of CHD received a diagnosis of CHD by 1 week of age, 88% by 1 year, and 99% by 4 years.

CYANOTIC CONDITIONS

Cyanosis is the term used to describe the bluish discoloration of the skin or mucous membranes as a result of inadequate oxygenation of the blood. The conditions that present as cyanotic babies or infants are those where there is right to left shunting of blood in the heart so that flow to the lungs is reduced or abnormal arterial connections and at the severe end of the spectrum hearts where there is only a single ventricular chamber. The cyanotic heart conditions include transposition of the great arteries, pulmonary atresia, truncus arteriosus, total anomalous pulmonary venous drainage, tetralogy of Fallot, and hearts with a single functional ventricle.

ACYANOTIC CONDITIONS

Acyanotic heart conditions do not cause reduced blood oxygen levels and a number of conditions lead to left to right shunting of blood within the heart due to septal defects resulting in excessive flow to the lungs. Affected children often present with breathlessness, which in turn leads to difficulty feeding and failure to thrive. In cases where there is a large left to right shunt the child will develop congestive cardiac failure. This group of conditions includes ventricular septal defect, atrioventricular septal defect, patent ductus arteriosus, and atrial septal defect. Other important acyanotic conditions are those where there is obstruction to the outflow of either the right or left side of the heart such as pulmonary valve stenosis, aortic valve stenosis, and coarctation of aorta.

LATER PRESENTATION OF CONGENITAL HEART DISEASE

Less serious CHD may not cause symptoms in infancy and may be detected incidentally by the finding of a heart murmur on routine examination. Lesions in this category include atrial septal defect, small ventricular septal defect, and mild aortic or pulmonary valve stenosis. Some of these conditions may not require intervention.

THE DIAGNOSTIC TOOLS

The gold standard in the diagnosis of CHD in newborns is echocardiography (ultrasound examination of the heart). Echocardiography first came into use during the early 1970s[6] and with the advent of two-dimensional echocardiography, and the later addition of pulsed wave, continuous wave, and color Doppler, and subsequent improvements in image quality, it is now possible to fully characterize the anatomy of even very complex congenital heart lesions in most patients. The development of echocardiography has led to early and accurate diagnosis in the majority of cases without resort to more invasive techniques such as angiography. In the current era the increasing use of cardiac MRI[7] and CT has helped to augment the diagnostic information available, and therefore the improving quality of detailed diagnosis in turn has a very significant positive influence in guiding and improving outcomes for surgical procedures.

CARDIAC CATHETERIZATION

Cardiac catheterization is a procedure where a very fine tube (catheter) is passed from a vein or artery usually at the groin up to the heart. The procedure is performed in a special theater under continuous X-ray guidance called fluoroscopy. In children cardiac catheterization is usually performed under general anesthesia. The purpose of the procedure may be:

1. Diagnostic, to gain more information about the anatomical abnormality of the heart and to measure pressures in the chambers and vessels. A radio-opaque contrast material is injected through the catheter to obtain detailed angiograms outlining the anatomy. In recent years diagnostic cardiac catheterization has been largely superseded by cardiac MRI.
2. Interventional, to treat the heart condition. This technique has been one of the most rapidly evolving areas in pediatric cardiology in the past 30 years.[8] It has become possible to treat many conditions, previously only treatable by surgery, with this less invasive technique. Using special balloon catheters valvar stenoses of either the pulmonary[9] or aortic valves[10] can be treated very successfully. Occlusion devices delivered through small catheters are now widely used to close atrial septal defects,[11] patent ductus arteriosus,[12] and some forms of ventricular septal defects.[13] In the past 10 years the placement of stents at interventional catheterization has been increasingly used to treat pulmonary arterial stenoses[14] and coarctation of aorta[15] in carefully selected cases. The ability to treat a wide range of conditions without using open surgery is a major advance avoiding creating a surgical scar and shorter hospital stays with quicker recovery for the patient.

SURGERY FOR CONGENITAL HEART DISEASE

The first reported surgery for a congenital heart defect was in 1939 when Gross and Hubbard[16] operated on a 7-year-old child to ligate a patent ductus arteriosus. In 1945 the first surgical procedure for a cyanotic heart condition was performed by Alfred Blalock when he carried out the first "shunt" operation in a child with tetralogy of Fallot.[17] This involved creating an anastomosis between the subclavian artery and pulmonary artery to improve the blood flow to the lungs.

A landmark development in the surgical treatment of CHD was the development of open heart surgery during the mid-1950s. This for the first time allowed intracardiac defects to be treated surgically. Open heart surgery was pioneered by an American surgeon Walton Lillehei who came to be regarded as the "Father of open heart surgery." In 1954 Lillehei and colleagues successfully used cross-circulation between a living human donor and the patient to perform open heart surgery and close a ventricular septal defect.[18]

CARDIOPULMONARY BYPASS

The technique of cardiopulmonary bypass (CPB) was developed by Gibbon and Kirklin, who pioneered its clinical use.[19] CPB is the technique in which the function of the heart and lungs is taken over during surgery by an oxygenator circuit and pump allowing the surgeon to work on the nonbeating heart. The progressive development and refinement of CPB now allows intracardiac defects such as atrial septal defect, ventricular septal defect, atrioventricular septal defect, tetralogy of Fallot, transposition of the great arteries, and other lesions to be repaired routinely using open heart surgery.

CPB is usually combined with hypothermia where the blood is cooled in the pump circuit and returned to the body. The cooled blood slows the body's basal metabolic rate, decreasing its demand for oxygen thus protecting the brain and vital organs. For work on the aortic arch *deep hypothermic circulatory arrest* may be necessary. It involves cooling the body to 12–18°C and stopping the blood flow. The body and the brain in particular can tolerate this for short period of time, and circulatory arrest beyond 30 min does increase the risk of some degree of permanent brain injury. In an attempt to protect the cerebral circulation during cardiac surgery the technique of *low-flow CPB* has been developed. This form of CPB maintains low-flow cerebral flow during surgery, and Newberger et al.[20] (as described in detail later in this book) have reported that infants having heart surgery using low-flow CPB had less early neurological morbidity as compared with those where the surgery was performed under predominantly circulatory arrest. In trying to define the key factors determining the neurodevelopmental outcome in children who have had surgery for CHD much attention has been focused on the refinement of the techniques of cerebral and myocardial protection during CPB.

The last 30 years have seen further great strides in the refinement of surgical techniques, combined with high-quality postoperative intensive care so that now the surgical repair of many cardiac defects has become routine. In the United Kingdom it is mandatory to report the outcome of all pediatric cardiac surgical procedures to a central database managed by the National Institute for Cardiovascular Outcomes Research (NICOR).[4]

Table 1.1 is an extract from the NICOR Congenital Heart Disease Website showing the survival rates during 2013–2014 for some of the most common major procedures illustrating the current excellent outcomes in children's heart surgery.

The transformation in outcomes for complex congenital heart disease is probably best illustrated by the condition of hypoplastic left heart syndrome (HLHS). Thirty years ago most children with this condition were offered no treatment and thus passed away after few days. Development of surgical procedures now means that the majority of babies born with HLHS do have treatment. This condition can now usually be treated by a series of three operations which reconstruct the circulation such that the right ventricle becomes the systemic ventricle pumping the blood around the body and the veins bringing the systemic venous return back to the heart are redirected such that deoxygenated blood flows passively to the pulmonary arteries. The surgeries are complex and they do not repair the defect but can allow the majority of those with HLHS to have a good quality of life in childhood. In the current era the survival rate through all three stages of the surgery for HLHS is greater than 75%. This new cohort of children surviving with complex CHD is an ever-increasing one and gives rise not only to a whole new population of patients with an ongoing need for expert medical input but also input from a multidisciplinary team. The long-term future for children with HLHS and other conditions where there is a single ventricle circulation is uncertain.

TABLE 1.1   Data from the National Institute for Cardiovascular Outcomes Research Congenital Heart Disease Audit Website, UK, 2013–2014 Indicate Excellent Survival to 30 days After Surgery

Operation	Outcome in UK Centers 2013–2014 (Survival at 30 days Postoperative)
Arterial switch repair for isolated transposition of great arteries	100%
Complete atrioventricular septal defect	99.5%
Fontan procedure	98.2%
Isolated coarctation repair	97.7%
Norwood procedure stage I	88.8%
Tetralogy of Fallot repair	99.7%
Ventricular septal defect repair	99.7%

In writing about this era Robert Anderson[20] made the following very insightful statement:

> The last 30 years of cardiac surgery have been like an exhilarating, breathless tour of the Himalayas with one "last great problem" after another succumbing to the surgeon's knife. Now we must address the irritating question of what has been achieved. What kind of life can be offered to these blue and breathless mites, snatched from the jaws of death?

This very much crystallizes the core theme of this book.

The operation is "corrective" for about 85% of those with CHD requiring surgery. The remaining 15% are those with complete CHD, mainly patients with a univentricular heart where either the left or right ventricle is hypoplastic. This group of patients can be surgically treated; however, the procedures are aimed at changing the circulation so that it can work without aiming to correct the defect. For most children the surgical program will involve a series of three operations eventually leading to a "Fontan" circulation where the deoxygenated blood from the superior and inferior vena cavae is redirected directly to the pulmonary arteries and the single ventricle which pumps blood to the systemic circulation.[21] This type of surgery is termed "palliative," and these complex conditions have an impact on long-term survival and the quality of life. Intuitively one might expect that those with complex CHD might be the children who suffer most from behavioral and psychological problems as a result of their cardiac problem, but as this book demonstrates that is not always the case. Many children with complex CHD show remarkable resilience and lead very full and productive lives.

SUCCESS IN THE TREATMENT OF CONGENITAL HEART DISEASE

The transformation of outcomes over the past generation has now shifted the focus for those teams caring for children with CHD to measure "success" in broader terms other than survival alone.

Achieving the above aims requires the input of many disciplines as illustrated in Fig. 1.1 and a philosophy in the team providing care of treating the "whole child." About 20% of children with CHD will have another significant comorbidity[22] emphasizing the need for all elements of the multidisciplinary team to contribute to care.

In our own Departments of Paediatric Cardiology and Clinical Psychology in Belfast we first began more than 20 years ago to study the outcomes for CHD in terms of physical, psychosocial, behavioral, and educational outcomes.[23,24] Since then we have had an ongoing research program including the Congenital Heart Disease Intervention Programme (CHIP).[25] That work and the knowledge gained from the studies of the other key authors in this book have helped us better understand the important

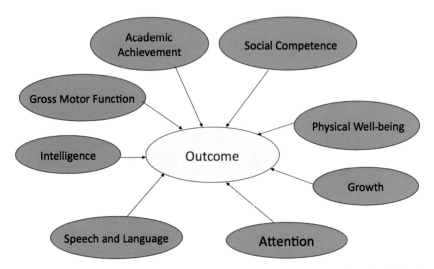

FIGURE 1.1   Key elements contributing to the "success" of the outcome for the child with congenital heart disease.

factors influencing the outcomes, and how some may be changed. This new knowledge is now applied in the ongoing clinical care of children and their families, which is the ultimate measure of success of any research.

The new era of detailed brain imagining by MRI is also an exciting opportunity to study the structural developments of the brain in CHD before and after birth. Our knowledge of the correlation between the structural differences in brain development in CHD with subsequent clinical outcomes is still in its infancy. It is, however, already clear that prenatal factors have an important role.

Study of the impact of, in particular, neonatal surgery involving CPB with, in some patients, circulatory arrest is already influencing the brain perfusion techniques that are used during surgery and also the intensive care management postoperatively.

The increasing body of research available to us already informs us that the ultimate developmental outcome for the child with CHD is potentially influenced at many points from fetal life through childhood as eloquently illustrated in Fig. 1.2 (courtesy of Hovels-Gurich et al.).

Our aim must therefore be to modify the risk factors where possible and provide treatment and support for those that are beyond our control.

ADULT CONGENITAL HEART DISEASE

The success story of children's heart surgery has given rise to a very large and new population of young adults with CHD who have survived childhood surgery. Many do not require further surgery, but it is important

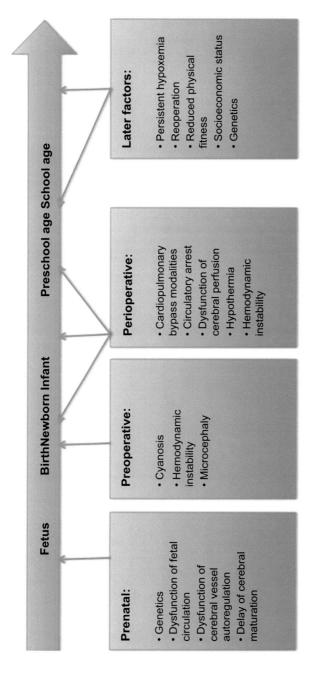

FIGURE 1.2 Time axis of risk factors for psychomotor development.

Adapted from: Hövels-Gürich HH: Psychomotor development of children with congenital heart defects. Causes, prevalence and prevention of developmental disorders after cardiac surgery in childhood. Monatsschr Kinderheilkd 2012; 160:118–128.

Herbera U and Hövels-Gürich HH: Neurological and Psychomotor development of foetuses and children with congenital heart disease–causes and prevalence of disorders and long-term prognosis. Z Geburtsh Neonatol 2012; 216:1–9.

to note that in the United States more patients over the age of 16 years have surgery each year for CHD than in the under 16 years age group. Many of these procedures are reoperations. For the first time we now have young adults with single ventricle hearts surviving to adulthood with the "Fontan" circulation. This presents to us medical issues not previously encountered, and in adulthood the single ventricle circulation can become more difficult to manage leading to an increasing physical and psychosocial impact on the life of the affected person. Within this group are those who need to be considered for heart transplantation. It should also be emphasized that one of the most gratifying outcomes in CHD is to see young adults who have come through all their treatment successfully and now leading active, productive lives including having their own children.

The story of treatment for CHD is therefore a very positive one, and there is scope to improve the "whole patient" approach to ensure that each child with congenital heart disease can achieve their maximum potential.

References

1. Neill CA. Etiology of congenital heart disease. In: Engle MA, ed. *Paediatric cardiology*. Cardiovascular Clinics; Vol. 4; 1972:138–147.
2. Ferencz C, Neill CA. Cardiovascular malformations: prevalence at birth. In: Freedom RM, Benson L, Smallhorn J, eds. *Neonatal heart disease*. 1st ed. London: Springer–Verlag; 1992:19–29.
3. Wilkinson J, Cooke WI. Cardiovascular disorders. In: Robertson NRC, ed. *Textbook of neonatology*. 2nd ed. London: Chuchill Livingstone; 1992:559–603.
4. National Institute for Cardiovascular Outcomes Research(NICOR). *Congenital heart disease web portal: Antenatal detection*; 2013–2014.
5. Hoffman JL, Kaplan S. The incidence of congenital heart disease. *J Am Coll Cardiol*. June 19, 2002;39(12):1890–1900.
6. Edler I, Lindstrom K. The history of echocardiography. *Ultrasound Med Biol*. December 2004;30(12):1565–1644.
7. Pohost GM. The history of cardiovascular magnetic resonance. *J Am Coll Cardiol Imaging*. 2008;1(5):672–678.
8. Mullins CE. History of pediatric interventional catheterizations. *Pediatr Cardiol*. January–February 1998;19(1):3–7.
9. Rao PS. Balloon pulmonary valvuloplasty: a review. *Clin Cardiol*. February 1989;12(2): 55–74.
10. Fratz S, Gildein HP, Balling G, et al. Aortic valvuloplasty in pediatric patients substantially postpones the need for aortic valve surgery. *Circulation*. 2008;117:1201–1206.
11. Molke L, Ramaciotti C. Atrial septal defect treatment options. *AACN Clin Issues*. April–June 2005;16(2):252–266.
12. Boehm W, Emmel M, Sreeram N. The Amplatzer duct occluder for PDA closure: indications, technique of implantation and clinical outcome. *Images Paediatr Cardiol*. April–June 2007;9(2).
13. Chen ZY, Lin BR, Chen WH, et al. Percutaneous device occlusion and minimally invasive surgical repair for perimembranous ventricular septal defect. *Ann Thorac Surg*. April 2014;97(4):1400–1406.
14. Holzer RJ, Gauvreau K, Kreutzer J, et al. Balloon angioplasty and stenting of branch pulmonary arteries. *Circ Cardiovasc Interv*. 2011;4:287–296.

15. Godart F. Intravascular stenting for the treatment of coarctation of the aorta in adolescent and adult patients. *Arch Cardiovasc Dis*. 2011;104:627–635.

16. Gross RE, Hubard JP. Surgical ligation of a patent ductus arteriosus; report of first successful case. *JAMA*. 1939;112:112729–112731.

17. Blalock A, Taussig H. The surgical treatment of malformations of the heart in which there is pulmonary stenosis or pulmonary atresia. *JAMA*. 1945;128:189–202.

18. Goor DA. *The genius of Walton Lillehei and the true story of open heart surgery*. 1st ed. New York: Vantage Press; 2007:305–307.

19. Kirklin JW. Open heart surgery at the Mayo Clinic - the 25th anniversary. *Mayo Clin Proc*. 1980;50:339–341.

20. Anderson RH. The fate of survivors of surgery for congenital heart disease. In: Anderson RH, Shinebourne EA, McCartney F, Tynan M, eds. *Paediatric cardiology*. Chuchill Livingstone Inc.; 1987:1363–1370.

21. D'Udeken Y, Lyenger A, Cochrane A, et al. The Fontan procedure: contemporary techniques have improved long-term outcomes. *Circulation*. 2007;116:I157–I164.

22. Massin MM, Astadicko I, Dessy H. Non cardiac co-morbidities of congenital heart disease in children. *Acta Paediatr*. May 2007;96(5):753–755.

23. Casey FA, Craig BG, Mulholland HC. Quality of life after surgical palliation for complex congenital heart disease. *Arch Dis Child*. 1994;70:382–386.

24. Casey FA, Sykes D, Craig BG, Power R, Mulholland HC. Behavioural adjustment of children with surgically palliated complex congenital heart disease. *J Pediatr Psychol*. 1996;3:335–352.

25. Mc Cusker CG, Doherty NN, Molloy B, et al. Determinants of neuropsychological and behavioural outcomes in early childhood survivors of congenital heart disease. *Arch Dis Child*. 2007;92:137–142.

2

Historical Perspectives in Pediatric Psychology and Congenital Heart Disease

N. Rooney

The Queen's University of Belfast, Belfast, United Kingdom

THE EMERGENCE OF PEDIATRIC MEDICINE

While there are many references to the medical treatment of children throughout ancient history, France led the way in establishing pediatric medicine as a specialism, when the first pediatric hospital in the world, the *Hôpital des Enfants Malades*, opened in Paris in 1802. It treated children up to the age of 15 years and continues to operate as a pediatric hospital, having merged in 1920 with the adult *Hôpital Necker*.

Gradually other European countries followed the trend of providing separate children's hospitals, including Berlin in 1830, St. Petersburg in

15

1834, and Vienna in 1837. Britain opened its first children's hospital *Great Ormond Street Hospital* exactly 50 years later in 1852, with the United States opening its first children's hospital in Philadelphia in 1855 and the famous *Boston Children's Hospital* opening its doors in 1869. This pattern of nursing and treating children separately has continued, leading to the establishment of children's hospitals throughout the world, many of which focus on particular pediatric medical and surgical specialisms.

However, it was Dr. Abraham Jacobi (1830–1919) who is now recognized as the founder of pediatrics. Born in Germany, he moved to New York where he opened the first Department of Pediatrics in a general hospital, while in England Dr. George Still (1868–1941), who worked in *Great Ormond Street Hospital*, was appointed the first Professor of Pediatrics. The section on *Pediatrics* of the *American Medical Association* was formed in 1880, although very few doctors exclusively practiced in the specialty before the start of the century. This picture changed as advances in the pasteurization of milk and the prevention of childhood rickets and scurvy through administration of cod liver oil and orange led more doctors to become interested in the prevention and treatment of childhood disease.[1] Despite this, it was not until 1953 in the United Kingdom that the *British Association of Paediatric Surgeons* (BAPS) was established, with the *Royal College of Paediatrics and Child Health* established as recently as 1996.

HISTORY OF CLINICAL PSYCHOLOGY

It is interesting to note similarities in the development of psychology, where although the study of the mind and behavior can be dated back to ancient times, psychology's foundations as an independent discipline in the 1890s can also be linked to an interplay between Germany and America. It was in Leipzig that Wilhelm Wundt established the first psychology research laboratory, referring to himself as a "psychologist," while Hermann Ebbinghaus (1850–1909) was pioneering the study of memory and Ivan Pavlov (1849–1936) was developing the model of classical conditioning which informs much of the practice of clinical psychology to this day.

In the United States developments in pragmatism and functional psychology by William James (1842–1910) and in education and scientific pedagogy by G. Stanley Hall (1844–1924) secured the establishment of psychology as a scientific profession.

Coinciding with this development of academic disciplines the *American Psychological Association* (APA) was founded in 1892 and consisted of a president, G. Stanley Hall (1844–1924), and 31 members. It expanded quickly following the Second World War, and now has 54 divisions in subspecialisms in psychology and 134,000 members.

England followed suit with the *British Psychological Society* being founded in 1901, initially as the professional body for teachers of psychology. Its purpose was *to advance scientific psychological research, and to further the cooperation of investigators in the various branches of psychology.* It currently has 10 divisions and 13 sections representing the various subspecialisms, the largest of these being the *Division of Clinical Psychology.* The latter is subdivided into faculties, the largest of which is the *Faculty for Children, Young People and Their Families.*

Viewed as the founder of "clinical psychology" Lightner Witmer (1867–1956) undoubtedly also laid the foundations of pediatric psychology.

Witmer introduced the term "clinical psychology" and opened the first psychology clinic at Pennsylvania University in 1896 with the aim of studying and treating children presenting with serious behavior and mood problems. He was a prolific researcher on childhood behavioral, emotional, and developmental problems. Witmer established the scientific journal *The Psychological Clinic* and much of the work then published was typically undertaken by child psychologists working with educational and learning difficulties.

THE CONVERGENCE OF CLINICAL PSYCHOLOGY AND MEDICINE

Until the early 20th century, applied psychology and medicine developed along their own professional trajectories. However, this was to change in 1910 when another coincidental link between Germany and the United States was to forge the relationship between medicine and psychology. The *Flexner Report*, published at this time, is generally viewed as the most important development in the history of American medical training.[2] Flexner introduced the then German model of medical training which encompassed the integration of the biomedical science model with experiential clinical training. In 1911, in response to the Flexner report's new scientific focus in medical training, the *APA* identified an opportunity to apply the scientific knowledge base of psychology within medicine. It established a committee to develop the relationship between psychology and medical training and surveyed US medical schools on their views of psychology in medicine. Responses were positive, recommending inclusion of psychology in the training curriculum for medical practitioners.[3]

While the seeds were sown, development was slow. The *Perkins Law* of 1915 allowed for crippled children of poor parents to access free medical treatment from the medical college hospital, which provided further impetus for specialist pediatric departments to be established. The development of research into the treatment of childhood diseases gained momentum and, importantly, it was recognized that other disciplines had much to contribute here. Interdisciplinary professional groups such as the *Society for Research in Child Development* emerged and began to generate much of the knowledge base to inform training and practice.[4]

When John Anderson, Professor of the *Institute of Child Welfare* at the University of Minnesota, made his presentation in 1930 at the *Convention of the American Medical Association* about greater collaboration between pediatrics and child psychology, he particularly highlighted the role of intelligence testing and parent training. This elicited a limited response at the time, and it was not until after the Second World War that federal funding was directed at employing psychologists in medical schools, although this had a largely psychiatric focus.

Nevertheless, as the number of specialist pediatricians grew over subsequent decades, so too did recognition of a need for parent training to manage children with developmental difficulties, chronic illness, and behavior problems. This resulted in J. L. Wilson, the president of the *American Pediatric Society*, proposing in 1964 in a presidential address that groups of pediatricians should actually hire clinical psychologists to work with them in solving the presenting problems of childhood and adolescence which they were facing.[5]

Conceptual models that gave a scientific basis to what clinicians were recognizing as pragmatic necessity began to emerge. Jerome Kagan[6], a professor of developmental psychology, proposed a *new marriage* of pediatrics and psychology in understanding the development of psychopathologies in the context of chronic illness and disability in childhood.[6] In so doing, psychologists' important roles in the early identification of disorders and in implementing interventions were highlighted. Logan Wright (1933–99), president of the APA, coined the term pediatric psychology, and wrote many conceptual papers on the specialism which he defined as *dealing primarily with children in a medical setting which is non-psychiatric in nature.*[7] Wright's view of a pediatric psychologist's role was based on a consultant scientist–practitioner model, involving the application of behavioral interventions and short-term therapy aimed at parent training and promoting child development—rather than one which was aimed at treating comorbid psychiatric disorders per se.

In 1967, with professional and indeed public momentum for pediatric psychology established, the *APA* appointed a committee to consider the emergent developments within pediatrics and psychology. In 1968 Logan Wright, along with child psychologists Lee Salk and Dorothea Ross, formed the *Society for Pediatric Psychology* (SPP). The society published a quarterly newsletter "Pediatric Psychology" from 1969 to 1975, which developed into the "Journal of Pediatric Psychology," edited by Donald Routh and Gary Mesibov. Today the SPP's mission statement is "… to promote the health and psychological well-being of children, youth and their families through science and an evidence-based approach to practice, education, training, advocacy and consultation." Today it has 1700 members and is composed of 17 special interest groups, which demonstrates the growth in specialisms within pediatric psychology.

PEDIATRIC PSYCHOLOGY: AN EVOLUTION IN RESEARCH AND CONCEPTS

Early psychological research in the area of child development provided an important context for developments in pediatric psychology beyond simply looking at parent training to manage behaviors. Such research was spearheaded by child analysts John Bowlby at the Tavistock Clinic in London[8] and by Rene Spitz in France.[9,10] This work focused on attachment and considered the psychological effects on children who had been separated from their parents and taken to places of safety during the London blitz and on children who had been institutionalized. Others applied this work to study the effects of separation on hospitalized children, reporting that psychological detriment resulted from such enforced separation of children from their mothers.[11,12]

This work demonstrated the importance of the mother–child relationship in promoting psychological resilience and normal development in the child. The research into the impact of hospitalization on child development extended beyond emotional development, with further early studies highlighting the negative outcomes on cognitive and attainment variables at school (eg, reading and social and communication competencies).[13,14]

Related research into child temperament highlighted a complex interplay between inherent child characteristics (including illness and disability) and environmental influences (notably the nature of the mother–infant relationship) in later child behavior and personality development.[15]

Thus when pediatric psychology was emergent, psychologists were strongly influenced by a recognition of the importance of the maternal role in managing children in hospital and through pediatric conditions of childhood. From early on, however, and in a way which would make psychologists distinct from their child psychiatry counterparts, an important role was advocated to *work through* medical and nursing colleagues. Thus, important interventions formulated were about educating doctors, nurses, and hospital administrators on the need to ensure children were not separated from parents during hospitalization, or when undergoing painful medical interventions, or how to go about this when such was unavoidable,[16,17] as well as emphasizing the rights and importance of children being appropriately informed about their illness and care.[18] Such interventions continue to be a major plank of interventions in pediatric psychology today.

However, other developments within society were having an impact on psychological research. The emergence of feminism during the 1970s led to research which challenged the established views that mothers working outside of the home negatively impacted on child development.[19] Such research identified the positive aspect of other caring relationships in the child's system and also highlighted the important role of fathers in child development.[20] Urie Bronfenbrenner's[21] ecological systems theory of human development was a particularly influential model put forward at this time. This highlighted the need to consider outcomes for the child within multiple systems of influence, namely, familial, peer networks, communities, health, education, and social services.[21] This challenged traditional models of simple causal relationships among illness, disability, and behavioral adjustment and at the same time started to broaden the possible foci for psychosocial interventions in pediatric psychology.

It would take some time for such conceptual models to translate into the applied evidence base. While such thinking was beginning to impact on pediatric research, the developing evidence base was very much focused on measuring long-term psychosocial outcomes for children with chronic illness, such as hemophilia,[22] cystic fibrosis,[23] and diabetes[24] within an essentially unidirectional causal model. Similar methodologies were inevitably applied within the field of congenital heart disease.

CONGENITAL HEART DISEASE AND PEDIATRIC PSYCHOLOGY

In 1936 the publication of *Abbott's Atlas of Congenital Heart Disease* put congenital heart disease on the radar of medical science by listing 1000 specimens of congenital cardiac abnormalities and linking them to the presence or absence of cyanosis.[25] In 1939 the first surgical treatment for a congenital abnormality was accomplished in *Boston Children's Hospital* by Dr. Robert Gross who successfully ligated a patent ductus arteriosus in a child.[26] In November 1944 history was made when the first "blue baby" surgery was successfully performed on Eileen Saxon, an extremely cyanotic infant due to her severe tetralogy of Fallot.[27] Shortly after this in 1947, Brooke Taussig (1898–1986), known as the "mother" of pediatric cardiology, authored *Congenital Malformations of the Heart*, a clinical text that documented congenital malformations and linked them to clinical and postmortem details.[27,28] It quickly became the leading text on medical and surgical management of congenital heart disease. Subsequently, medical and surgical interventions in congenital heart disease developed expeditiously, largely due to the collaboration of pediatric cardiologists and cardiac surgeons.

In 1957 the *American Academy of Pediatrics* established the subspecialty section of *Pediatric Cardiology*. International development of this specialty followed in 1964 when the *Association for European Paediatric Cardiology* was formed, with 1980 seeing the first World Congress of Paedriatric Cardiology and Cardiac Surgery held in London.

With the improvement in survival rates for children with congenital heart problems the condition was now perceived as a chronic disease and thus became another focus of research for pediatric psychologists. The impact of the illness on childhood development was generally measured through the identification of psychological sequelae such as psychosocial and cognitive deficits and general quality of life.

Early studies of children with congenital heart disease had suggested that their intellectual development was usually within the normal range, although typically at the lower range,[29,30] with some degree of cognitive impairment noted in older children with cyanotic congenital heart disease in particular.[31,32] Although hypoxia was the proposed explanation even these early studies suggested that parental "overprotection" might inhibit access to normal developmental activities,[33] with timing of surgical repair being an additional contributory factor.[34]

Subsequent early studies pursued these efforts to map out and understand the psychological nature of the cognitive and behavioral difficulties faced by these childhood survivors of congenital heart disease. Intellectual and cognitive outcomes were most widely assessed using the *Wechsler Intelligence Scales for Children*,[35] while studies focusing on behavioral outcomes typically utilized parental assessments of their child's behavior: for example, the *Child Behavior Checklist* (CBCL).[36] This parental measure involved an assessment of the child's overall behavior and emotion in terms "internalizing" (eg, depression, anxiety, and withdrawal) and "externalizing" (eg, hyperactivity and aggression) problems.

However, the research results have reported varied findings in terms of the impact of the disease on cognitive and psychosocial development. Several studies have identified increased behavioral problems, with higher levels of internalizing problems, such as social withdrawal or somatic symptoms in older children, and externalizing behaviors, such as attention deficit hyperactivity disorder and impulsivity, more prevalent in younger children with heart disease.[37–39] Early studies noted that there appeared to be some relationship between cognitive deficits and behavioral outcomes.[39] Other studies, however, did not find group differences when these children were compared to control groups or test norms and some even reported more favorable outcomes.[40–42]

As McCusker and Casey discuss in chapter "Is There a Behavioral Phenotype for Children with Congenital Heart Disease?," there is potential for much "noise" in the pediatric psychology literature, and it takes time for consistent findings and patterns to become apparent. Although De Maso alerted us as early as 1991, for example, that maternal perceptions of disease severity may be at odds with disease severity per se—but may be a better predictor of outcome[43]—too often research has confounded outcomes reporting with respondent factors influencing the same. More studies were needed which would systematically control for, or attend to, the impact of this and other important factors such as variability in the type and severity of the congenital heart disease, the misinterpretation of physical symptoms as emotional features in behavioral rating scales, developmental considerations in sampling, small cohort number, and the reliance on retrospective versus prospective designs. Most importantly, a central theme of this book is that research designs in this area need to be based on the sort of *systemic* model of understanding child outcomes such as that advocated by Bronfenbrenner. As is outlined in the later chapters of this book, research into understanding outcomes for children with congenital heart disease has indeed become informed in this way. In particular, the importance of family functioning, as a factor at once affected by the congenital heart disease and in turn as an important predictor of outcome therein, has informed the work reported in these chapters. As outlined in the following chapters, cognitive and behavioral phenotypes are now emerging. Moreover, a clearer understanding of risk and protective factors across disease, surgical, congenital, and environmental domains is also not emergent which importantly illuminates potential pathways to secondary prevention and treatment.

The next important step, however, has indeed been to translate this growing understanding into the development of effective psychological interventions. A special edition of the *Journal of Pediatric Psychology* in 2014 noted the significant progress that has been made during the past decade in demonstrating the effectiveness of specific psychological interventions for children with chronic illnesses and their families.[44] The *Congenital Heart disease Intervention Program* (CHIP), to be described in the later chapters of this book, represents the first such major trial across the world for children with congenital heart disease. Exemplifying Jerome Kagan's *marriage* of psychology and pediatrics, in a secondary prevention program targeted at key developmental transitions, the CHIP project highlights that today we are at an exciting juncture, where our children with congenital heart disease can indeed benefit from the knowledge that this cross-discipline collaboration brings, in leading fuller, happier, and more effective lives.

References

1. Richmond JB. Child development: a basic science for pediatrics. *Pediatrics*. 1967;39:649–658.
2. Moll W. History of American medical education. *Br J Med Educ*. 1968;2:173–181.
3. Fernberger SW. The American psychological association: a historical summary, 1892–1930. *Psychol Bull*. 1932;29:1–89.
4. Routh DK. The short history of pediatric psychology. *J Pediatr Psychol*. 1975;4:6–8.
5. Wilson JL. Growth and development of pediatrics. *J Pediatr*. 1964;65:984–991.
6. Kagan J. The new marriage: pediatrics and psychology. *Am J Dis Child*. 1965;110:272–278.

7. Wright L. The pediatric psychologist: a role model. *Am Psychol*. 1967;22:323–325.
8. Bowlby J. Maternal care and mental health. *Bull World Health Organ*. 1951;3:355–534.
9. Spitz RA. Hospitalism: an inquiry into the genesis of psychiatric conditions in early childhood. *Psychoanal Study Child*. 1945;1:53–74.
10. Spitz RA, Wolf KM. Anaclitic depression. *Psychoanal Study Child*. 1947;2:313–342.
11. Prugh D, Staub EM, Sands HH, Kirschbaum RM, Lenihan EA. A study of the emotional reactions of children and families to hospitalization and illness. *Am J Orthopsychiatry*. 1953;23:70–106.
12. Robertson J. Some responses of young children to loss of maternal care. *Nurs Times*. 1953;49:382–386.
13. Douglas JW. Early hospital admissions and later disturbances of behaviour and learning. *Dev Med Child Neurol*. 1975;17:456–480.
14. Quinton D, Rutter M. Early hospital admissions and later disturbances of behaviour: an attempted replication of Douglas' findings. *Dev Med Child Neurol*. 1976;18:447–459.
15. Thomas A, Chess S. *Temperament and Development*. New York: Brunner/Mazel; 1977.
16. Waechter EH. Children's awareness of fatal illness. *Am J Nurs*. 1971;71:1168–1172.
17. Seagull EAW. The child's rights as a medical patient. *J Clin Child Psychol*. 1978;7:202–205.
18. Koocher GB. Talking with children about death. *Am J Orthopsychiatry*. 1974;44:404–411.
19. Hoffman LW. Maternal employment. *Am Psychol*. 1979;34:859–865.
20. Lamb ME. *The Role of the Father in Child Development*. New York: Wiley; 1976.
21. Bronfenbrenner U. *The Ecology of Human Development: Experiments by Nature and Design*. Cambridge: Harvard University Press; 1979.
22. Markova I, Sterling-Phillips J, Forbes CD. The use of tools by children with haemophilia. *J Child Psychol Psychiatr Allied Discipl*. 1984;25:261–272.
23. Stark LJ, Jelalian E, Miller DL. Cystic fibrosis. In: Roberts MC, ed. *Handbook of Pediatric Psychology*. 2nd ed. New York: Guildford Press; 1995.
24. Anderson BJ, Laffel LM. Behavioral and psychosocial research with school-aged children with type 1 diabetes. *Diabetes Spect*. 1997;10:23–28.
25. Abbott ME. *Atlas of Congenital Cardiac Disease*. New York: The American Heart Association; 1936.
26. Gross RE, Hubbard JP. Surgical closure of a patent ductus arteriosus: report of a first successful case. *JAMA*. 1939;112:729–731.
27. Kothari SS. A 'sense' of history and pediatric cardiology. *Ann Pediatr Cardiol*. 2011;4:1–2.
28. Taussig HB. *Congenital Malformations of the Heart*. New York: The Commonwealth Fund; 1947.
29. Chazan M, Harris T, O'Neill D, Campbell M. The intellectual and emotional development of children with congenital heart disease. *Guys Hosp Rep*. 1951:331–341.
30. Kramer HH, Awiszus D, Sterzel U, Van Halteren A, Claben R. Development of personality and intelligence in children with congenital heart disease. *J Child Psychol Psychiatry*. 1989;30:299–308.
31. Feldt RH, Ewert JC, Stickler GB, Weidman WH. Children with congenital heart disease: motor development and intelligence. *Am J Dis Child*. 1969;117:281–287.
32. Linde LM, Rasof B, Dunn OJ. Longitudinal studies of intellectual and behavioral development in children with congenital heart disease. *Acta Paediatr Scand*. 1970;59:169–176.
33. Linde LM, Rasof B, Dunn OJ. Mental development in congenital heart disease. *J Pediatr*. 1967;71:198–203.
34. Newburger JW, Tucker AD, Silbert AR, Flyer DC. Motor function and timing of surgery in transposition of the great arteries, intact ventricular septum. *Pediatr Cardiol*. 1983;4:317.
35. Wechsler D. *Wechsler Intelligence Scale for Children – Fourth Edition, UK (WISC-IV)*. Psych-Corp, Harcourt Assessment; 2003.
36. Achenbach TM, Rescorla LA. *Manual for the ASEBA School-Age Forms and Profiles*. Burlington, VT: University of Vermont, Research Center for Children, Youth and Families; 2001.

37. Casey FA, Sykes DH, Craig BG, Power R, Mulholland HC. Behavioural adjustment of children with surgically palliated complex congenital heart disease. *J Pediatr Psychol.* 1996;21:335–352.

38. Oates RK, Turnbull JA, Simpson JM, Cartmill TB. Parent and teacher perceptions of child behaviour following cardiac surgery. *Acta Paediatr.* 1994;83:1303–1307. 214.

39. Utens EM, Verhulst FC, Meijboom FJ, et al. Behavioural and emotional problems in children and adolescents with congenital heart disease. *Psychol Med.* 1993;23:415–424.

40. Karsdorp P, Everaerd W, Kindt M, Mulder B. Psychological and cognitive functioning in children and adolescents with congenital heart disease: a meta-analysis. *J Pediatr Psychol.* 2007;32(5):527–541.

41. Utens E, Verslusis-Den Bieman H, Witsenburg M, Bogers AJ, Verhulst FC, Hess J. Cognitive and behavioural and emotional functioning of young children awaiting elective cardiac surgery or catheter intervention. *Cardiol Young.* 2001;11:153–160.

42. Salzer-Mohar U, Herle M, Floquet P, et al. Self-concept in male and female adolescents with congenital heart disease. *Clin Pediatr.* 2002;41:17–24.

43. DeMaso DR, Campis LK, Wypij D, Bertram S, Lipshita M, Freed M. The impact of maternal perceptions and medical severity on the adjustment of children with congenital heart disease. *J Pediatr Psychol.* 1991;16:137–149.

44. Palermo TM. Evidence based interventions in pediatric psychology: progress over the decades. *J Pediatr Psychol.* 2014;39:753–762.

PART II

TOWARD A NEURO-DEVELOPMENTAL PHENOTYPE

A Longitudinal Study From Infancy to Adolescence of the Neurodevelopmental Phenotype Associated With d-Transposition of the Great Arteries

D.C. Bellinger

Harvard Medical School and Boston Children's Hospital, Boston, MA, United States; Harvard School of Public Health, Boston, MA, United States

J.W. Newburger

Harvard Medical School and Boston Children's Hospital, Boston, MA, United States

OUTLINE

Our group has conducted research on the neurodevelopment of children with critical congenital heart disease (CCHD) since the late 1980s. Some studies have been randomized trials, comparing the outcomes associated with different intraoperative strategies, while others have been cross-sectional studies comparing the outcomes of children with various heart defects to those of normative populations or a referent sample. In this chapter, we summarize the major findings of the Boston Circulatory Arrest Study (BCAS). We followed up children with dextro-transposition (d-transposition) of the great arteries (d-TGA) from before the arterial switch operation was performed through hospital discharge, and assessed neurodevelopment when the children were 1, 2½, 4, 8, and 16 years of age. The results of earlier assessments suggested hypotheses about the children's relative weaknesses that we used to design later assessments. In this way, over time we developed an increasingly detailed appreciation of the neurodevelopmental phenotype of children with d-TGA.

The BCAS was designed to compare the neurodevelopmental outcomes associated with two vital organ support strategies: deep hypothermia with either predominant total circulatory arrest or predominant continuous low-flow cardiopulmonary bypass.[1] Although we were able to draw inferences about the length of total circulatory arrest that would not be expected to result in adverse outcomes,[2] the outcomes in the two groups were more striking in their similarities than their differences. Therefore, we here focus on findings common to both treatment groups rather than on treatment group differences.

1-YEAR ASSESSMENT

The most prominent early finding was a modest deficit in motor development, with 20% of the children achieving a Psychomotor Development Index score (on the 1969 version of the *Bayley Scales of Infant Development*) of 80 or below.[3] Item analyses revealed that fewer children than expected passed age-appropriate motor milestones, such as "neat pincer" and "pat-a-cake: midline skill."[4] The infants were also less vocal than expected, suggesting the possibility of a delay in expressive language development.

2½-YEAR ASSESSMENT

We obtained parents' ratings of the children's language development using the *MacArthur Communicative Development Inventory*.[4] Delays of 2–4 months were found on all subscales (*Vocabulary Production, Word Use, Word Endings, Irregular Forms, Overregularizations, Sentence Complexity, and Mean Length of Utterance*). Six percent of the children did not yet produce two-word utterances (vs none in the standardization sample). On most scales of the *Child Behavior Checklist/2–3*, parents reported significantly fewer problems than did parents of children in the standardization sample. Children not yet combining words were rated less optimally on both the *Internalizing* and *Externalizing* Behavior scales, however.

4-YEAR ASSESSMENT

On a broad-based battery of assessments, the children performed below population norms in several areas, including expressive language, visual–motor integration, and gross and fine motor function.[5] Although the Full-Scale IQ score of most children was within the average range, the variability in their scores on the components of the *Wechsler Preschool and Primary Scale of Intelligence-Revised* suggested that such apical scores (ie, those that integrate performance over multiple diverse tasks) are somewhat misleading. The children showed marked weaknesses on subtests that assessed visual–spatial and visual–motor integration skills.

An assessment by a speech and language pathologist revealed that, despite normal oral musculature and normal hearing, many children had oral–motor coordination and speech planning problems. These were expressed as difficulty imitating oral movements and speech sounds (eg, "stick out your tongue" and repeat "pa-ta-ka" rapidly). Their speech was marked by a variety of phonological deviations, including cluster reductions and simplifications, omission of medial and final consonants, and sequencing errors. These reduced the intelligibility of their speech, especially when contextual cues were not available as an aid to the listener. Almost 25% of children met diagnostic criteria for apraxia of speech.

It was apparent that something else was amiss with the children's language development, though it did not pertain to the size of their vocabularies or the syntactic complexity of their sentences. Rather it involved higher-order linguistic skills, specifically pragmatics or the use of language in social situations. With the assistance of Lowry Hemphill, a psycholinguist from the Harvard Graduate School of Education, we began to administer two tasks that assess narrative discourse skills. These were elicited personal narratives, in which the child was provided prompts and models and asked to tell three stories about events that most children have experienced (eg, going to the beach, being stung by a bee, and having a spilling accident), and a free play session with a parent using a standard set of toys. In contrast to typically developing children, children with d-TGA had great difficulty generating personal narratives, with a substantial percentage being unable to produce even one (10% vs 0% in the referents).[6] Moreover, the narratives they did produce were poorer in construction, failing to orient the listener in terms of time and place or to convey a logical sequence of actions. In free play, the child and parent engaged in fewer symbolic play episodes, which were defined as sequences of at least two logically linked actions, with a specified setting, participants with an intention, conflict, and resolution (eg, taking a bath and going to sleep).[7] Although the children with d-TGA produced as much language as the referents during the interaction, it tended to be more primitive. While referents took a leading role in constructing and elaborating play episodes, playing character roles and narrating action the speech of children with d-TGA consisted primarily of object labeling and sound effects, leaving the parent to provide most of the verbal scaffolding for the interaction.

8-YEAR ASSESSMENT

At this assessment, we focused on pragmatic language, visual–spatial skills, and both fine (including oral–motor coordination) and gross motor function. Once again, most children achieved scores that fell within normal limits, but below the expected level in many domains, including academic achievement, fine motor function, visual–spatial skills, working memory, hypothesis-generating and testing, sustained attention, and higher-order language skills.[8]

In general, lower-level skills within a domain were relatively intact, but the children had difficulty integrating or coordinating them to achieve higher-order goals. For example, although most children scored at age level on a test of single word reading, many scored lower than expected on a test of reading comprehension, which involves extracting meaning from the connected discourse. Similarly, many children had acquired basic mathematical concepts but had difficulty applying them to solve specific computational problems. As would be expected, the children were at increased risk of academic struggles, and more than one-third of them had already received remedial academic services.

Several findings suggested that executive functioning is a domain that is particularly vulnerable, especially the management of meta-cognitive dimensions of behavior, such as planning and organization, which underlie performance on diverse tasks. This was evident on both nonverbal and verbal tasks. On the *Rey–Osterrieth Complex Figure*, the percentage of children whose copy was scored at the lowest level of organization was more than twice the percentage observed in the standardization sample.[9] They focused on the details at the expense of the figure's strong organizational elements, with the result that the copies were poorly formed. We conducted a "clinical sort," classifying children's copies into five categories based on the extent to which the overall organization of the figure was preserved. We found a monotonic, "dose-related," relationship between sort category and the percentage of children receiving remedial assistance in school. More than half of the children who produced the poorest copies were receiving assistance, compared to 10% of the children who produced the best copies.

Executive dysfunction was also evident in the children's oral and written higher-order language skills. Written narratives produced in response to a standard three-picture prompt were more likely than those of referents to omit reference to key events that provided the story with its temporal coherence.[10] Elicited personal narratives at age 8, like those at age 4, did not adequately address the information needs of a listener, resembling the narratives produced by typically developing children who are several years younger. For both the figure copying and pragmatic language tasks, good performance rests on the ability to perceive the overall organization of a set of data, and then to plan, structure, monitor, and modify output based on feedback. Whether the task involved the assembly of story elements into a coherent narrative or the accurate representation of a complex design, the children with d-TGA appeared to become lost in the details, failing to see how the pieces fit to make a well-formed whole.

In contrast to previous assessments, both parents and teachers identified behavioral vulnerabilities. The children had elevated frequencies of both internalizing and externalizing problem behaviors, as well as poorer competence (ie, parent ratings of activities, social adaptation, and school performance) and adaptive skills (ie, teacher ratings of academic performance, effort, appropriateness of behavior, learning, and affect).[11] Children who had previously shown signs of CNS disturbance (eg, ictal activity in the postoperative period, abnormal neurological findings at age 4, and lower IQ at age 4) were at particular risk. Moreover, children experiencing academic difficulties in school at age 8 showed a greater shift toward disordered behavior between the ages of 4 and 8 years than did children not experiencing academic difficulties. These findings suggested that the children would be at increased risk of psychiatric morbidity in adolescence. This concern was also based on their executive function (eg, working memory) and pragmatic language difficulties. From the observed neurocognitive testing results, we hypothesized that BCAS subjects were at risk for deficits in social cognition,[12] which concerns the ability to appreciate and make effective use of knowledge about the mental states of other people, such as identifying another's emotional state and understanding his or her perspective and information needs.

16-YEAR ASSESSMENT

We focused on executive functions, memory, social cognition, and psychiatric status.[13] As at prior assessments, most of the patients performed within the average range, especially on apical indices. However, the distributions of test scores were generally shifted toward poorer scores, with increased percentages of children falling in the range of clinical concern. Moreover, within a domain, certain aspects proved to be more of a challenge than others. For instance, on the *Children's Memory Scale* (Table 3.1), adolescents scored near the expected mean values on story recall (both immediate and delayed) and on recitation of overlearned sequences, but lower than the expected means on recalling faces and on tasks that involved learning new material, whether verbal or visual, that lacked the helpful structure of stories (namely, the location of dots in a matrix, arbitrary word pairs, and random number strings). Interestingly, on both dot locations and word pairs, scores tended to be higher on the trials that involved recall following a delay than on initial learning trials. This suggests that they registered the material initially, had trouble organizing its retrieval in the short term, but were able to do so having had some time to consolidate it. This could reflect generalized anxiety, uncertainty about task demands, or a tendency to become overwhelmed, at least initially, due to slow information processing.

The adolescents continued to struggle with aspects of executive functioning. The nine subtests of the *Delis–Kaplan Executive Function System* were administered. The scores were very close to the expected mean values on all aspects of only three subtests (*Design Fluency, Twenty Questions,* and *Tower*) (Table 3.2). On the other six, they had difficulty on

TABLE 3.1   Scores of Adolescents With d-TGA on the Children's Memory Scale

Subtest	Mean (SD) (Expected 10, 3)
Dot Locations	
Learning	8.3 (3.6)
Long delay	9.4 (2.9)
Stories	
Immediate	10.2 (3.1)
Delayed	9.9 (2.9)
Faces	
Immediate	8.4 (3.4)
Delayed	8.9 (3.1)
Word Pairs	
Learning	7.3 (3.5)
Long delay	8.3 (3.3)
Numbers	7.9 (3.3)
Sequences	9.6 (2.9)

TABLE 3.2   Scores of Adolescents With d-TGA on the Delis–Kaplan Executive Function System

Subtest	Mean (SD) (Expected 10, 3)
TRAIL-MAKING TEST	
Visual Scanning	10.0 (2.6)
Number Sequencing	9.1 (3.1)
Letter Sequencing	9.4 (3.2)
Number–Letter Sequencing	8.3 (3.3)
Motor Speed	11.2 (2.4)
VERBAL FLUENCY	
Letter Fluency	8.6 (3.5)
Category Fluency	9.8 (3.1)
Category Switching: Total correct	8.6 (3.3)
DESIGN FLUENCY	
Filled Dots	10.3 (3.0)
Empty Dots	10.0 (3.1)
Switching: Total correct	10.4 (3.0)
COLOR–WORD INTERFERENCE TEST	
Color Naming	8.3 (3.2)
Word Reading	9.2 (3.4)
Inhibition	8.1 (3.6)
Inhibition/Switching	8.5 (3.4)
SORTING TEST	
Free Sorting	
Correct sorts	7.9 (2.6)
Description	7.9 (2.7)
Sort Recognition	6.8 (3.3)
TWENTY QUESTIONS TEST	
Initial Abstraction	10.4 (3.1)
Weighted Achievement	10.2 (3.4)
WORD CONTEXT TEST	
Total consecutively correct	8.7 (3.2)

Continued

TABLE 3.2   Scores of Adolescents With d-TGA on the Delis–Kaplan Executive Function System—cont'd

Subtest	Mean (SD) (Expected 10, 3)
TOWER TEST	
Total achievement	9.7 (2.5)
PROVERBS TEST	
Total achievement free inquiry	8.6 (3.5)
Accuracy	8.0 (3.6)
Abstraction	7.6 (3.9)

at least some trials. On the *Trail-Making Test*, they did well on the simpler trials of *Visual Scanning, Number Sequencing, Letter Sequencing*, and *Motor Speed*, but significantly worse on *Number–Letter Switching*, particularly when compared with their score on *Motor Speed*. The component skills (eg, number sequencing and alphabetic sequencing) were intact, but they had difficulty switching back and forth between these well-learned sequences, suggesting cognitive inflexibility. Similarly, on *Verbal Fluency*, while they were much more proficient at *Category Fluency* (ie, naming objects in specified semantic categories) than they were in *Letter Fluency* (ie, retrieving words that begin with specified letters), they had marked difficulty on the trial of *Category Fluency* that required them to alternate between categories (ie, name a fruit, then a piece of furniture, and then a fruit). This result also suggested that cognitive flexibility is a relative weakness. On the *Color–Word Interference Test*, the adolescents did better on *Word Reading* (ie, reading words that consist of color names) than they did on *Color Naming* (ie, identifying the colors of patches) or on the most complex trials of *Inhibition* and *Inhibition/Switching*. In this Stroop-like task, the *Inhibition* trial required them to ignore salient cues and name the color of the ink in which a word was printed when that color conflicted with the word (eg, the word "red" was printed in blue ink). The adolescents' scores on *Inhibition/Switching* were actually somewhat higher than those on *Inhibition*, despite the fact that the former trial would appear to be more difficult, involving switching response sets (ie, naming the color of ink in which a word was printed, except when the word was enclosed in a box, in which case, the word was to be read). This finding, like some of those on the delay trials of subtests of the *Children's Memory Scale*, suggests that once the adolescents had some experience with a task, their performance improved substantially (ie, more than would be expected simply from practice). Again, one possibility is that initial uncertainty or anxiety about what was expected of them interfered with their performance on early trials, but that, with time, they were able to regroup and perform well.

The *Sorting* test was the subtest on which the adolescents had the greatest difficulty. The adolescents had difficulty sorting the cards themselves, but even greater difficulty on the *Sort Recognition* trial. This was the trial in which the examiner sorted the cards and asked the adolescent to identify the principle used. Often an adolescent was unable to identify a sorting principle that he or she had applied just minutes before. Lower scores on this subtest also suggest cognitive inflexibility (ie, difficulty thinking about the same material in different ways) and the ability to engage in abstract thinking. On *Word Context*, adolescents were asked to identify a target word relying on a series of verbal clues. Adolescents with d-TGA required more clues to deduce the target, suggesting a relative weakness in abstraction and the ability to integrate multiple pieces of information to converge on a conclusion. On *Proverbs*, the adolescents had difficulty providing the meanings of familiar and unfamiliar proverbs, consistent with a reduced ability to decode complex language constructions and nonliteral meaning.

We asked parents, teachers, and the adolescents themselves to complete the *Behavior Rating Inventory of Executive Function*, which provides information about executive functioning in natural settings (Table 3.3). In general, the adolescents' self-ratings were in line with population values. In contrast, the scores provided by parents were as much as ½ standard deviation higher (worse), with the higher scores generally being on the scales contributing to the meta-cognition index (*Initiate, Working Memory, Plan/Organize, Organization of Materials*, and *Monitor*). The scores provided by teachers were even more elevated, exceeding the expected values by as much as a standard deviation. One of three scores contributing to the *Behavioral Regulation Index* (*Shift*), and three of five scores contributing to the *Meta-Cognition Index* (*Initiate, Working Memory*, and *Plan/Organize*) showed an elevation of 1 standard deviation or greater. The standard deviations tended to exceed the expected value for parent and, especially, teacher ratings, suggesting the presence of adolescents with rather extreme scores.

We explored the adolescents' social cognition skills using the "*Reading the Mind in the Eyes" Test-Revised*, as well as two self-report questionnaires, the *Autism Screening Quotient*, and the *Toronto Alexithymia Scale*. Compared to referents, the adolescents with d-TGA were less accurate in identifying the emotions of the models in the pictures, scored significantly higher in terms of the *Autism Screening Quotient* (indicating more autistic traits), and significantly worse on parts of the *Toronto Alexithymia Scale*, indicating an inability to focus on their own emotional experience and emotional states.

Detailed assessments of psychiatric status and history were carried out using the *Kiddie-Schedule of Affective Disorders* and parent- or self-completed questionnaires that provided dimensional assessments of depression, anxiety, and disruptive behavior.[14] Compared with the

TABLE 3.3   Scores of Adolescents With d-TGA on the Behavior Rating Inventory of Executive Function: Self-Report, Parent, Teacher Ratings

	Mean (SD) (Expected 50, 10)
SELF-REPORT	
Behavioral Regulation Index	50.0 (11.9)
Inhibit	50.1 (11.8)
Shift	49.7 (11.7)
Emotional Control	49.2 (10.7)
Meta-Cognition Index	51.4 (11.5)
Monitor	50.4 (11.2)
Working Memory	51.7 (11.6)
Plan/Organize	51.1 (11.6)
Organization of Materials	50.9 (10.2)
Task Completion	51.6 (11.6)
Global Executive Composite	50.8 (11.8)
PARENT REPORT	
Behavioral Regulation Index	51.5 (11.8)
Inhibit	50.9 (10.8)
Shift	52.4 (12.0)
Emotional Control	51.1 (11.2)
Meta-Cognition Index	56.1 (12.1)
Initiate	54.5 (12.7)
Working Memory	56.0 (14.1)
Plan/Organize	55.5 (12.1)
Organization of Materials	54.2 (10.1)
Monitor	56.1 (11.5)
Global Executive Composite	54.9 (12.2)
TEACHER REPORT	
Behavioral Regulation Index	55.6 (13.6)
Inhibit	52.1 (10.4)
Shift	60.3 (20.5)
Emotional Control	53.0 (13.1)

TABLE 3.3   Scores of Adolescents With d-TGA on the Behavior Rating Inventory of Executive Function: Self-Report, Parent, Teacher Ratings—cont'd

	Mean (SD) (Expected 50, 10)
Meta-Cognition Index	60.9 (17.0)
Initiate	60.0 (16.6)
Working Memory	60.6 (16.8)
Plan/Organize	60.1 (16.1)
Organization of Materials	57.9 (18.1)
Monitor	57.7 (13.6)
Global Executive Composite	60.3 (16.5)

referent group, adolescents with d-TGA had significantly higher lifetime prevalence of any psychiatric disorder (42% vs 25%), with the difference largely due to a higher frequency of ADHD in the d-TGA group (19% vs 7%). Use of stimulants (20% vs 3%) and antidepressants (11% vs 0%) was significantly more common in the TGA patients. Clinicians assigned the TGA patients significantly lower global psychosocial functioning scores on the *Children's Global Assessment Scale* and on the *Brief Psychiatric Rating Scale for Children*. On the *Children's Depression Inventory*, adolescents with d-TGA had higher scores on the *Interpersonal Problems* and *Ineffectiveness* scales (and total score). On the *Revised Children's Manifest Anxiety Scale*, they scored worse on the *Physiological Anxiety* and *Social Concerns/Concentration* scales (and total score). Parents of the TGA adolescents, as well as the adolescents themselves, reported significantly more ADHD-related behaviors on all subscales of the *Connors' ADHD/DM-IV* scale (total score, *Hyperactive–Impulsive*, *Inattentive*, and *ADHD Index*), while teachers reported significantly higher scores only on the *Inattentive* subscale. On the *Conduct Disorder Scale*, the adolescents with d-TGA reported significantly higher total score (*Conduct Disorder Quotient*), as well as on the subscales *Aggressive Conduct*, *Deceitfulness/Theft*, and *Rule Violation*. Finally, on the *Child Stress Disorders Checklist-Screening Form*, these adolescents scored significantly higher on the *Numbing and Dissociation*, *Increased Arousal*, and *Impairment in Functioning* subscales (as well as total). Impaired cognitive functioning and parental stress at age 8 years were significant risk factors for psychiatric dysfunction at 16 years.

We hypothesize that the neuropsychological deficits of children with d-TGA make establishing and maintaining relationships challenging, perhaps contributing to increased psychiatric risk. Adolescent peer interactions are often rapid in pace, involving telegraphic references to people and events, requiring rapid information processing. They are rich in affect, requiring rapid and accurate interpretation of facial expressions and body language, and thus good social cognitive skills. Allegiances shift rapidly, needing to be updated regularly to avoid social blunders, requiring good cognitive flexibility and working memory. The language is complex and indirect, involving constructions whose meaning cannot be deduced from linguistic cues alone (eg, sarcasm and innuendo), and thus require strong pragmatic language skills. Similarly, managing the frequent peer conflicts that arise requires that one be able to perceive and interpret pertinent external and internal cues, identify a desired goal, evaluate alternative strategies for achieving it, implement the strategy selected, and revise it as needed. The neuropsychological phenotype we have described for these children would likely make peer interactions fraught with risk.

Although the frequency of ADHD is frequently noted to be elevated among children with CCHD, we hypothesize that a more appropriate diagnosis would be "sluggish cognitive tempo" (SCT).[15] Such children are described as "easily confused," "spacey or in a fog," and "doesn't process information as quickly or as accurately as others." The social dysfunction and comorbidities of children with SCT contrasts with that of children with ADHD, but resembles that observed among children with CCHD, involving social withdrawal, isolation, and depression.

At 16 years of age, structural neuroimaging showed that the frequency of "any abnormality" was eightfold greater among the adolescents with TGA than the referents (33% vs 4%).[13] The abnormalities tended to be focal (23%) rather than diffuse (3%), with mineralization or iron deposition being the most common focal findings. Focal atrophy or infarction was found in 6% of TGA patients. Although evidence of gross white matter injuries was not found, diffusion tensor imaging revealed disturbances in white matter microstructure, with significantly reduced fractional anisotropy (FA) in the deep white matter of the cerebral hemispheres (frontal, parietal, and temporal lobes), cerebellum, and pontomesencephalic region, particularly in subcortical and periventricular regions.[16] Significant correlations were found between FA values in different regions and adolescents' scores on neuropsychological tests (mathematics achievement and inattention/hyperactivity symptoms with left parietal FA; inattention/hyperactivity symptoms and executive function with right precentral FA; visual–spatial skills with right frontal FA; and memory and right posterior limb of the internal capsule).[17]

In conclusion, in the BCAS, while we showed that children with d-TGA manifest signs of neurodevelopmental dysfunction early in life, the early assessments failed to reveal many of the vulnerabilities that were eventually revealed. For many children, their struggles increased rather dramatically in severity when they entered the late primary grades. We believe this is because they were unable to meet the increasing demands placed on their higher-order executive function skills at this stage of schooling and beyond. In the early years, when parents and then teachers do much of the planning and organization that structures children's lives, the children's limitations were not so apparent, but the eventual transfer of these responsibilities to the children exposed their vulnerabilities. We furthermore hypothesize that the neuropsychological weaknesses also impact the children's social functioning and their risk for psychiatric disorders. The BCAS cohort has now reached adulthood (age >25 years); future studies will ascertain whether neurocognitive impairments in adolescence portend risk of lower educational attainment, unemployment, and absence of meaningful relationships in the adult with CHD.

References

1. Newburger JW, Jonas RA, Wernovsky G, et al. A comparison of the perioperative neurologic effects of hypothermic circulatory arrest versus low-flow cardiopulmonary bypass in infant heart surgery. *N Engl J Med*. 1993;329(15):1057–1064.
2. Wypij D, Newburger JW, Rappaport LA, et al. The effect of duration of deep hypothermic circulatory arrest in infant heart surgery on late neurodevelopment: the Boston circulatory arrest trial. *J Thorac Cardiovasc Surg*. 2003;126(5):1397–1403.
3. Bellinger DC, Jonas RA, Rappaport LA, et al. Developmental and neurologic status of children after heart surgery with hypothermic circulatory arrest or low-flow cardiopulmonary bypass. *N Engl J Med*. 1995;332(9):549–555.
4. Bellinger DC, Rappaport LA, Wypij D, Wernovsky G, Newburger JW. Patterns of developmental dysfunction after surgery during infancy to correct transposition of the great arteries. *J Dev Behav Pediatr*. 1997;18(2):75–83.
5. Bellinger DC, Wypij D, Kuban KC, et al. Developmental and neurological status of children at 4 years of age after heart surgery with hypothermic circulatory arrest or low-flow cardiopulmonary bypass. *Circulation*. 1999;100(5):526–532.
6. Hemphill L, Uccelli P, Winner K, Chang C-J, Bellinger D. Narrative discourse in young children with histories of early corrective heart surgery. *J Speech Lang Hear Res*. 2002;45:318–331.
7. Ovadia R, Hemphill L, Winner K, Bellinger D. Just pretend: participation in symbolic talk by children with histories of early corrective heart surgery. *Appl Psycholinguistics*. 2000;21:321–340.
8. Bellinger DC, Wypij D, duPlessis AJ, et al. Neurodevelopmental status at eight years in children with dextro-transposition of the great arteries: the Boston circulatory arrest trial. *J Thorac Cardiovasc Surg*. 2003;126(5):1385–1396.
9. Bellinger DC, Bernstein JH, Kirkwood MW, Rappaport LA, Newburger JW. Visual-spatial skills in children after open-heart surgery. *J Dev Behav Pediatr*. 2003;24(3):169–179.
10. Beck SW, Coker D, Hemphill L, Bellinger D. Literacy skills of children with a history of early corrective heart surgery. In: Hoffman J, Schallert D, Fairbanks C, Maloch B, eds. *The 51st National Reading Conference Yearbook. Oak Creek, WI: National Reading Conference.* 2002:106–116.
11. Bellinger DC, Newburger JW, Wypij D, Kuban KC, duPlesssis AJ, Rappaport LA. Behaviour at eight years in children with surgically corrected transposition: The Boston circulatory arrest trial. *Cardiol Young*. 2009;19(1):86–97.
12. Bellinger DC. Are children with congenital cardiac malformations at increased risk of deficits in social cognition? *Cardiol Young*. 2008;18(1):3–9.
13. Bellinger DC, Wypij D, Rivkin MJ, et al. Adolescents with d-transposition of the great arteries corrected with the arterial switch procedure: neuropsychological assessment and structural brain imaging. *Circulation*. 2011;124(12):1361–1369.

14. DeMaso DR, Labella M, Taylor GA, et al. Psychiatric disorders and function in adolescents with d-transposition of the great arteries. *J Pediatr*. 2014;165(4):760–766.
15. Barkley RA. Distinguishing sluggish cognitive tempo from ADHD in children and adolescents: executive functioning, impairment, and comorbidity. *J Clin Child Adolesc Psychol*. 2013;42(2):161–173.
16. Rivkin MJ, Watson CG, Scoppettuolo LA, et al. Adolescents with D-transposition of the great arteries repaired in early infancy demonstrate reduced white matter microstructure associated with clinical risk factors. *J Thorac Cardiovasc Surg*. 2013;146(3):543–549.
17. Rollins CK, Watson CG, Asaro LA, et al. White matter microstructure and cognition in adolescents with congenital heart disease. *J Pediatr*. 2014;165(5):936–944.

Neurodevelopmental Patterns in Congenital Heart Disease Across Childhood: Longitudinal Studies From Europe

H.H. Hövels-Gürich

Aachen University of Technology, Aachen, Germany

C. McCusker

The Queens University of Belfast, Belfast, United Kingdom; The Royal Belfast Hospital for Sick Children, Belfast, United Kingdom

OUTLINE

The *Boston Circulatory Arrest Study* (BCAS) described in the Chapter 3 represents to date the most comprehensive prospective study of neurodevelopmental outcomes in children with congenital heart disease (CHD). An important story is emerging. Although motor and language difficulties are evident on early neurodevelopmental examination, more detailed assessments of later childhood suggest that more complex problems with integrative and executive functioning processes are emergent which may in fact underlie deficits on specific tasks of visuomotor, speech, attention, memory, and social cognition.

Although seminal, the BCAS involved only children with dextro-transposition of the great arteries (TGA)—a complex, usually corrected, cyanotic heart condition (see chapter: Congenital Heart Disease: The Evolution of Diagnosis, Treatments, and Outcomes). In this chapter longitudinal studies from Germany and Northern Ireland are described which involve patient samples with a range of other CHDs, both cyanotic and acyanotic, corrected and palliated only.

THE AACHEN TRIAL OF SCHOOL-AGE CHILDREN WITH TETRALOGY OF FALLOT AND VENTRICULAR SEPTAL DEFECT

This was a prospective trial of neurodevelopmental and behavioral outcomes in 40 school-age children (5–11 years) after biventricular corrective cardiac surgery in infancy. The sample selected did not have concomitant genetic or syndrome malformations. While all children had undergone low-flow cardiopulmonary bypass (CPB) and deep-hypothermic cardiac arrest—surgical processes which have been implicated in neurodevelopmental outcomes—distinctly we had two subgroups. These included a *cyanotic group* of children with tetralogy of Fallot (TOF) and preoperative hypoxemia (arterial oxygen saturation <90%) and an *acyanotic group* with a ventricular septal defect (VSD), heart failure, and pulmonary hypertension. All had undergone surgery in the first year of life (between 1993 and 1999) and socioeconomic status was not different between subgroups or in comparison to the normal population.

In 2004, patients underwent an extensive battery of standardized neurodevelopmental examinations, the results of which were compared to those of the normal population and between the subgroups.[1-4] Prospectively registered risk factors of preoperative, perioperative, and postoperative status and management were correlated with neurodevelopmental outcome parameters.

Positively, the quality of life in both clinical groups, and based on both child and parent reports, appeared comparable to peer-referenced norms. Moreover, and remarkably, exercise capacity was not different compared to the normal children and between the subgroups. However, similar to the TGA patients in the Boston study, a range of mild to moderate neuropsychological and psychological difficulties was apparent and especially in the cyanotic (TOF) subgroup:

- Decreases in general intelligence (IQ) were found for the whole group as well as for the cyanotic subgroup, whereas it was in the normal range for the acyanotic subgroup.

- Both the cyanotic and acyanotic subgroups were compromised on direct and parent-reported measures of academic achievement.
- Gross motor function was decreased in the total group, with a higher dysfunction rate in cyanotic compared to acyanotic children.
- Expressive and receptive language was reduced in the whole group and impairment was higher in cyanotic compared to acyanotic patients.
- Disturbances to specific aspects of the attentional network of executive functioning (monitoring and "resolving conflict") were evident in the cyanotic, but not the acyanotic subgroup. Other aspects of attention (alerting and orienting) were in the normal range in both subgroups.

There was an association between executive attentional dysfunction, motor dysfunction, reduced academic achievement, and increased risk for psychosocial maladjustment. Risk for psychological adjustment difficulties was elevated in the whole group without difference between the subgroups.

Thus, in our trial we confirmed that reduced neurodevelopmental outcomes are apparent in school-age children after CPB surgery in infancy, for both cyanotic and acyanotic conditions, with a stronger impairment in children with a cyanotic condition. In addition to this specific risk factor of preoperative hypoxemia, and socioeconomic status, and as in the Boston trials, we found that perioperative management factors, especially prolonged duration of CPB, increased the risk for neurodevelopmental impairment.[1-4]

THE BELFAST STUDIES

In 2001, and during preparation for the first early intervention trial of its kind (see chapters: The Congenital Heart Disease Intervention Programme (CHIP) and Interventions in Infancy; Growing Up – Interventions in Childhood and CHIP-School; and Healthy Teenagers and Adults – an Activity Intervention), we started to evaluate two cohorts of infants (<1 year) and children (4–5 years) with a range of congenital heart defects severe enough to warrant invasive surgical interventions in infancy. Our study populations consisted of children with both cyanotic and acyanotic conditions—those who underwent CPB procedures as well as transcatheter corrections only and those who had both corrective and palliative surgeries.

As in the Aachen trial, and when we excluded those children with genetic syndromes such as Down syndrome, we detected a greater range of mild to moderate deficits, across a range of cognitive abilities (eg, language, visuospatial, reasoning, and number skills), in our 4- to 5-year-old cohort ($n = 90$) with cyanotic compared to acyanotic conditions, and regardless of repair status.[5] Importantly, however, we found significant problems with sensorimotor functions across all clinical subgroups, irrespective of cyanosis, surgical procedure, or repair status. This is an important finding which challenges historical assumptions that neurological involvement is wholly related to the secondary impact of CPB and/or insufficient oxygenation arising from a cyanotic condition. It suggests that concomitant neurological and cardiological processes may be at play, and indeed since 2010 this has been the focus of enquiry and model building.[6,7]

Pervasive and significant deficits in perceptual–motor skills, relative to other early cognitive abilities (eg, memory and attention), were also noted in our infant cohort (average age = 8 months) about 6 months after surgery.[8] Disturbances in perceptual–motor functioning has not only been the most salient cognitive deficit found in these of our studies, but has also been the most consistently reported deficit in at least infants, and young children, with CHD, in the literature more generally.[9–11] Such a profile would appear to be a good candidate for the phenotype. However, as we followed up our infant sample across time, a more complex picture is emerging, comparable to what has been reported in the Boston trials.

Insult to the rapidly developing brain should not be expected to have a static effect. On the one hand, plasticity of function may result in recovery of early impairments. On the other hand, it is now well recognized that "late effects" may emerge across time as injured parts of the brain fail to develop properly and/or impairment becomes more apparent as later, more complex, skills fail to be acquired effectively.[12] The literature on children with CHD has been mixed, with some studies suggesting that deficits reduce with age,[9,13] and others suggesting that new deficits become apparent across time.[14]

We were able to follow up 31/54 of our original infant sample (excluding those with genetic conditions such as Down syndrome) 7 years after their original neuropsychological examination in the first year of life.[15,16] These now 7- to 8-year-olds underwent a comprehensive neuropsychological examination involving a range of visuospatial, sensorimotor, language, memory and reasoning abilities, and school attainments (and behavioral outcomes reported elsewhere in this volume).

Table 4.1 summarizes findings across three subgroups (children with corrected cyanotic conditions, acyanotic conditions, and with more complex, palliated defects). First, it is important to note that although there were many below-average profile scores, none actually fell in the clinically defective range (>2 standard deviations (SD) below the mean). Secondly, and unlike in our earlier study with 4- to 5-year-olds noted above, this time we did not see an appreciable difference between those with cyanotic and acyanotic conditions. Rather, taking 0.5–1 SD as indicative of a medium to large effect size,[17] both acyanotic and cyanotic corrected groups showed risk on 5 subscales with the palliated group showing risk

TABLE 4.1 Mean Standardized Scores Across Neurodevelopmental Domains in Children With Corrected Acyanotic Conditions, Cyanotic Conditions, and Palliated, Complex, Cyanotic Conditions

Cognitive domains	Acyanotic (*n* = 9)	Cyanotic (*n* = 9)	Complex (*n* = 13)
VERBAL REASONING			
Similarities[a]	10.7 (2.6)	10.2 (4.3)	10.2 (2.4)
NON-VERBAL REASONING			
Matrix reasoning[a]	8.6 (2.7)	7.8* (2.8)	7.8* (2.5)
LANGUAGE			
BPVS[b]	100.8 (8.2)	100.7 (6.8)	99.2 (13.1)
ATTENTION			
Digit span[c]	8.4* (3.6)	8.5* (2.7)	8.5* (3.1)
Auditory attention[c]	8.3* (4.7)	10.0 (3.5)	7.8* (4.3)
MEMORY			
Faces—immediate[c]	8.3* (1.9)	9.4 (2.9)	8.6 (2.5)
Faces—delayed[c]	10.6 (2.7)	9.8 (3.0)	9.0 (2.2)
Narrative memory[c]	8.4* (3.2)	8.0* (3.9)	8.5* (1.8)
SENSORIMOTOR SKILLS			
Imitating hand positions[c]	8.3* (2.6)	7.9* (2.8)	7.4** (3.9)
PERCEPTUAL–MOTOR SPEED			
Coding[a]	9.5 (2.9)	7.6** (3.0)	8.2* (3.3)
PERCEPTUAL–MOTOR SKILLS			
Visuomotor precision[c]	9.4 (4.4)	8.6 (3.3)	8.2* (3.1)
Design copy[c]	13.6 (2.2)	10.7 (2.4)	10.2 (5.2)

* Medium effect size.
** Large effect size.[17].
[a] *WISC-IV[19] subscales (expected mean = 10; SD = 3).*
[b] *Dunn LM, Dunn LM, Whetton C, Barley J.* British Picture Vocabulary Scale–II. *Windsor, England: NFER-Nelson, 1997 (expected mean = 100; SD = 15).*
[c] *NEPSY–II20 subscales (expected mean = 10; SD = 3).*

on 7 subscales. All groups appeared normal on tasks related to receptive language and verbal reasoning and all, regardless of cyanosis or correction status, appeared at risk in relation to processes of sensorimotor skills, attention, and narrative memory. Academic performance was compromised and the requirement for remedial input was elevated across these groups in comparison to sibling controls.[16]

Finally, we looked at concordance rates of those children who had been in the "at risk" range (>0.8 SD) on the perceptual–motor index and the mental development index subscales of the *Bayley Infant Scales of Development (BIDS-II)*[18] in infancy and the age-appropriate counterparts of these subscales at 7 years of age (taken from *WISC-IV.UK*[19] and *NEPSY-II*[20]). Specific concordance rates varied from as low as 48% (chance) up to 82%. However, the pattern was clear:

- Some recovery in the perceptual–motor deficits of infancy was evident. While 84% of infants were in the risk range in the first year of life, this figure had reduced to 39–42% at 7 years (as indicated by at risk scores on *any* of the perceptual–motor scales from the *WISC-IV.UK* and the *NEPSY-II*).
- However, in contrast, while only 19% of children had been in the at risk range on other aspects of cognitive functioning in infancy (as indicated by the *Mental Development Index* profile on the *BIDS-II*), this figure had increased to 42–67% as indicated by comparable scores on scales specifically related to attention, memory, and nonverbal reasoning at 7 years of age.

Thus, we see both some recovery in those cognitive deficits most apparent in infancy (perceptual–motor skills) and the emergence of difficulties in other aspects of cognitive functioning related to attention, memory, and nonverbal reasoning. A very similar pattern has also been observed at the Wisconsin Center in serial assessments of cognitive functioning across the first 3 years of life in children with a range of congenital heart defects.[21]

The question occurs then does this represent two different, partially independent, processes varying in parallel across time? Or, are disturbances with a common neurological basis being manifested in an artifactually different way depending on developmentally different manifestations of the same phenomena? This question is considered below.

FACTORS MEDIATING RISK

In both the Aachen and Belfast trials, as with the Boston series, perioperative factors (eg, cardiac arrest and duration of CPB) were found to be important risk factors for cognitive deficit.[1–5,15,16] This association has been manifested both on proximal outcomes evaluated close in time to surgery and on more distal outcomes evaluated many years later. This suggests robustness in this association, which has also been reliably noted in studies from other centers,[21–23] together with covarying factors such as length of time in cardiac recovery and in hospital per se.

Some studies, including our Aachen series, demonstrate an increased risk for neurodevelopmental disturbance to be determined by preoperative hypoxemia (thus implicating the cyanotic heart conditions more than the acyanotic; see chapter: Congenital Heart Disease: The Evolution of Diagnosis, Treatments, and Outcomes). MRI studies in 2010 and 2014 on long-term brain volume and macrostructure have demonstrated a marked reduction of total white and gray matter volumes in cyanotic more than in acyanotic adolescents, and in 21% of the total group an abnormal macrostructure was assessed. Both correlated with increased cognitive, motor, and executive dysfunctions.[23,24] This observation has been confirmed by a study on adolescents after infant surgery for TOF, 42% of whom presented with macrostructural brain abnormalities.[25] Related, CPB has not been shown to be related to decline beyond the infant age as observed in the cyanotic conditions.[26–28]

However, this latter association between the cyanotic conditions and outcomes appears more equivocal in the wider literature. In our Belfast studies we did find such an association in some of our studies, and for some cognitive domains, but not for others (see above). Thus children with both cyanotic and acyanotic conditions appeared at risk for sensorimotor deficits in infancy and early childhood and for higher-order problems in later childhood. Studies from some other centers are consistent with this with some even finding that children with acyanotic conditions are at greater risk than those with cyanotic conditions.[26,29,30] It would increasingly appear, at least, that if the cardiac defect is significant enough to warrant surgical intervention, then it is significant enough to increase neurodevelopmental risk, regardless of whether there has been preoperative hypoxemia or not.

Reasons for these sometimes contradictory findings in relation to cyanosis and outcomes have perhaps been suggested by recent studies into the primary neurological features of children with congenital heart disease (see below). These features may not only exert independent effects but also be interacting with the secondary impact of the heart defect itself and surgical processes on neurological functioning.

Growing evidence has emerged, for example, of prenatal and preoperative neurological abnormalities which are likely to be contributing to postoperative neurodevelopmental impairments. Intrauterine delay of brain maturation has been observed in fetuses, and after birth, in children with congenital heart disease in MRI studies[6,31,32] and correlations have been found with impaired neurobehavioral outcomes in mature neonates before cardiac surgery.[33] This brain immaturity, which putatively may vary across different specific heart conditions (regardless of the cyanotic classification) may render these groups particularly vulnerable to the additional impact of cyanosis and surgical processes and lead to intraoperative hypoxic–ischemic injury as well as to peri- and postoperative hemodynamic instability, leading to increased focal white matter injury.[7,34] The impact of congenital heart disease, for example, on fetal cerebrovascular blood flow dynamics has been suggested with respect to the delay of brain maturity and subsequent neurodevelopmental outcomes.[35,36] The findings in 2014, including evidence of abnormal intrauterine development of axons and myelin structures leading to permanent developmental deficits into adolescent age,[37] have culminated in a neurological model of an *encephalopathy of congenital heart disease* being proposed.[38]

Together, these interacting risk factors have implications for interventions. Potentially modifiable clinical risk factors such as perioperative cerebral oxygen delivery and postoperative cerebral blood flow velocity may be important foci here given their association with brain MRI abnormalities and indeed 1-year psychomotor outcomes in cyanotic and acyanotic infants.[39,40] However, also of significance for interventions, including those to be described later in this volume, is increased attention in the literature to environmental and family factors.

Although in our Belfast studies we found the greatest association to be with behavioral adjustment (see chapter: Is There a Behavioural Phenotype for Children With Congenital Heart Disease?), parenting style and maternal mental health and worry nonetheless also had significant predictive value for neurodevelopmental and school outcomes.[5,8,15,16] Moreover, as is described in chapter "The Congenital Heart Disease Intervention Programme (CHIP) and Interventions in Infancy", mother–infant interactions aimed at neurodevelopmental stimulation appeared to have beneficial effects, at least in the short term in dyads who received this training intervention compared to those who did not.[8] Associations beyond socioeconomic status and related to parental stress, family functioning, and cognitive outcome have indeed been noted in other centers.[26,29] Some, indeed have proposed that motor deficits may be just as likely due to compromised abilities to "practice" such skills early in infancy (eg, through reduced movement capacity and/or overprotection in a sick and fragile infant) as to primary neurological deficits per se.[41] Given that we now know that parenting style and especially maternal mental health, can impact significantly on infant brain development and cognitive outcomes in the wider population,[42] the import here for children with congenital heart disease should not be surprising.

CONCLUSIONS AND COMMENT

Findings from the Aachen and Belfast studies, with a range of different congenital heart conditions, appear to be telling a story remarkably similar to that of the Boston trials with children with TGA. Early in infancy and childhood, motor deficits (eg, sensorimotor, visuospatial, and speech articulation) appear to be preponderant. However, as longitudinal studies across child development have increased, it has become apparent that while some apparent recovery in these specific deficits occurs (on tests related to these specific functions), difficulties in other domains, not apparent in early childhood are emergent—attention, narrative memory, language, and higher intellectual processes. These findings are consistent with meta-analyses of earlier studies,[9,11] and with studies from other centers (in 2010 and 2014) involving older children.[10,26] Although when we exclude children with genetic syndromes such as Down syndrome, the extent of these deficits are relatively mild to moderate in nature, our research suggests that there is a distinct association between the extent of such deficits and academic and school outcomes.[15,16] Therefore, these findings are of more than academic interest; however, they have direct clinical, personal, and social implications.

The etiology of such a neurodevelopmental phenotype raises two possibilities. First, as noted above, it is possible that we are seeing some recovery across time in the neurological substrates of motor functioning—either through spontaneous plasticity of functional arrangements or because of the reason that over time these children get to effectively remediate these difficulties through "practice", as their physical capacities become less restricted and parental "overprotection" lessens. In parallel, we may be seeing a "growing into the deficit,"[12,26] or the emergence of late effects when other damaged parts of the brain fail to develop optimally, but the impact of which only becomes apparent in later childhood when such skills are demanded, and when pertinent neuropsychological tests can be used.

Alternatively, it is possible that there is a common neurological basis in operation, which is simply manifested in different ways across development. Thus, executive functioning skills of sequencing, integration, and filtering of information and aspects of attentional processing may be required for perceptual–motor functioning in early childhood before the latter become more automated and primarily controlled by parietal and nondominant hemisphere regions. The *Rey Complex Figure Test*,[43] for example, is normally a test implicating visuospatial operations in the nondominant hemisphere in adulthood. However, in childhood it is used as a test of the executive functions of planning and sequencing.[44] Thus, examination of the specific nature of the early (perceptual–motor) and later (circumscribed aspects of memory, language, social cognition, attentional, and nonverbal reasoning) deficits, which this corpus of work has suggested, proposes that integrative cognitive processes normally associated with executive functioning may be at play. As work turns more toward unpacking the cognitive processes underlying "outputs" on a range of neuropsychological tests,[41] together with more sophisticated neuroimaging of neural pathways involved, understanding of these possible etiologies is likely to advance.

A four-factor model of risk best encapsulates current understanding. As discussed above this includes (1) neurological features concomitant to, but partially independent of, the congenital heart defect, and possibly based on a delayed intrauterine brain maturation; (2) peri- and postoperative management; (3) preoperative hypoxemia; and (4) family and environmental factors. Also, it is possible that in addition to each exerting independent effect there will be interactions and bidirectional influences as suggested in Fig. 4.1. Thus concomitant neurological features may render the fetus and infant at particular susceptibility to adverse intrauterine and perinatal processes related to preoperative hypoxemia and perioperative processes. Environmental and family factors are, however, likely to exert an independent impact which either moderates against, or augments, adverse outcomes.

In Chapter 5 Marino synthesizes the work from these North American and European trials, in combination with work from elsewhere, and discusses the implications of this CHD phenotype for understanding future school and later adult outcomes. Most importantly, the consistency of the research evidence base is now such that clear implications for clinical practice, secondary prevention and interventions are warranted and indeed the *American Heart Federation* endorsed a consensus statement to this effect in 2012.[45] Implementation of this is considered and future research directions are proposed in the next chapters.

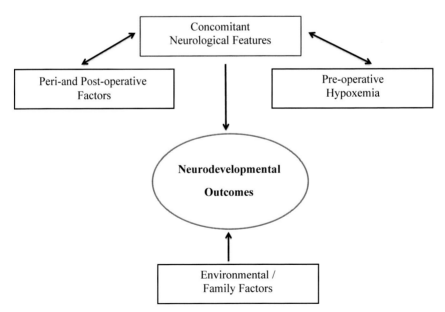

FIGURE 4.1   A four-factor model of risk for adverse neurodevelopmental outcomes in infants and children with congenital heart disease.

References

1. Hövels-Gürich HH, Konrad K, Skorzenski D, et al. Long-term neurodevelopmental outcome and exercise capacity after corrective surgery for tetralogy of Fallot or ventricular septal defect in infancy. *Ann Thorac Surg.* 2006;81(3):958–966.
2. Hövels-Gürich HH, Bauer SB, Schnitker R, et al. Long-term outcome of speech and language in children after corrective surgery for cyanotic or acyanotic cardiac defects in infancy. *Eur J Paediatr Neurol.* 2008;12(5):378–386.
3. Hövels-Gürich HH, Konrad K, Skorzenski D, Herpertz-Dahlmann B, Messmer BJ, Seghaye MC. Attentional dysfunction in children after corrective cardiac surgery in infancy. *Ann Thorac Surg.* 2007;83(4):1425–1430.
4. Hövels-Gürich HH, Konrad K, Skorzenski D, et al. Long-term behavior and quality of life after corrective cardiac surgery in infancy for tetralogy of Fallot or ventricular septal defect. *Pediatr Cardiol.* 2007;28(5):346–354.
5. McCusker CG, Doherty NN, Molloy B, et al. Determinants of neuropsychological and behavioral outcomes in early childhood survivors of congenital heart disease. *Arch Dis Child.* 2007;92:137–141.
6. Khalil A, Suff N, Thilaganathan B, Hurrell A, Cooper D, Carvalho JS. Brain abnormalities and neurodevelopmental delay in congenital heart disease: systematic review and meta-analysis. *Ultrasound Obstet Gynecol.* 2014;43(1):14–24.
7. McQuillen PS, Miller SP. Congenital heart disease and brain development. *Ann N Y Acad Sci.* 2010;1184:68–86.
8. McCusker CG, Doherty NN, Molloy B, et al. A controlled trial of early interventions to promote maternal adjustment and development in infants with severe congenital heart disease. *Child Care Health Dev.* 2009;36(1):110–117.
9. Karsdorp P, Everaerd W, Kindt M, Mulder B. Psychological and cognitive functioning in children and adolescents with congenital heart disease: a meta-analysis. *J Ped Psychol.* 2007;32(5):527–541.
10. Simons J, Glidden R, Sheslow D, Pizarro C. Intermediate neurodevelopmental outcome after repair of ventricular septal defect. *Ann Thorac Surg.* 2010;90:1586–1592.
11. Snookes SH, Gunn JK, Eldridge BJ, et al. A systematic review of motor and cognitive outcomes after early surgery for congenital heart disease. *Pediatrics.* 2010;125(4): e818–e827.
12. Anderson V, Dpencer-Smith M, Wood A. Do children really recover better? Neurobehavioural plasticity after early brain insult. *Brain.* 2011;134:2197–2221.
13. Menahem S, Poulakis Z, Prior M. Children subjected to cardiac surgery for congenital heart disease. Part 1 – Emotional and psychological outcomes. *Interact Cardiovasc Thorac Surg.* 2008;7:600–604.
14. McGrath E, Wypij D, Rappaport L, Newburger J, Bellinger D. Prediction of IQ and achievement at age 8 years from neurodevelopmental status at age 1 year in children with D-transposition of the great arteries. *Pediatrics.* 2004;114(5):572–576.
15. McCusker CG. Recovery –v- emergence of neurodevelopmental deficits in congenital heart disease from infancy to 7 years. In: *Paper presented at the 2nd Annual Cardiac Neurodevelopmental Symposium, Cincinnati, USA*; 2013.
16. McCusker CG, Armstrong MP, Mullen M, Doherty NN, Casey F. A sibling-controlled prospective study of outcomes at home and school in children with severe congenital heart disease. *Cardiol Young.* 2013;23(4):507–516.
17. Cohen JW. *Statistical power analysis for the behavioural sciences.* 2nd ed. Hillsdale, NJ: Lawrence Erlbaum Associates; 1988.
18. Bayley N. *Bayley scales of infant development.* 2nd ed. San Antonio, TX: The Psychological Corporation, Harcourt Brace & Company; 1993.
19. Wechsler D. *Wechsler intelligence scale for children.* 4th ed. UK (WISC-IV): PsychCorp, Harcourt Assessment; 2003.

20. Korkman M, Kirk U, Kemp SL. *NEPSY-II: A developmental neuropsychological assessment.* San Antonio, TX: Psychological Corporation; 2007.
21. Mussatto K, Hoffmann R, Hoffman G, et al. Risk and prevalence of developmental delay in young children with congenital heart disease. *Pediatrics.* 2014;133(3):e570–e577.
22. Mulkey SB, Swearingen CJ, Melguizo MS, et al. Academic proficiency in children after early congenital heart disease surgery. *Pediatr Cardiol.* 2014;35(2):344–352.
23. von Rhein M, Scheer I, Loenneker T, Huber R, Knirsch W, Latal B. Structural brain lesions in adolescents with congenital heart disease. *J Pediatr.* 2011;158(6):984–989.
24. von Rhein M, Buchmann A, Hagmann C, et al. Brain volumes predict neurodevelopment in adolescents after surgery for congenital heart disease. *Brain.* 2014;137(Pt 1):268–276.
25. Bellinger DC, Rivkin MJ, Demaso D, et al. Adolescents with tetralogy of Fallot: neuropsychological assessment and structural brain imaging. *Cardiol Young.* 2014;11:1–10.
26. Sarrechia I, De Wolf D, Miatton M, et al. Neurodevelopment and behavior after transcatheter versus surgical closure of secundum type atrial septal defect. *J Pediatr.* 2014;166(1):31–38. [Epub ahead of print].
27. Quartermain MD, Ittenbach RF, Flynn TB, et al. Neuropsychological status in children after repair of acyanotic congenital heart disease. *Pediatrics.* 2010;126(2):e351–e359.
28. van der Rijken R, Hulstijn-Dirkmaat G, Kraaimaat F, et al. Open-heart surgery at school age does not affect neurocognitive functioning. *Eur Heart J.* 2008;29(21):2681–2688.
29. Majnemer A, Limperopoulos C, Shevell M, Rohlicek C, Rosenblatt B, Tchervenkov C. Developmental and functional outcomes at school entry in children with congenital heart defects. *J Pediatr.* 2008;153:55–60.
30. Majnemer A, Limperopoulous C, Shevell M, Rohlicek C, Rosenblatt B, Tchervenkov C. A new look at outcomes of infants with congenital heart disease. *Ped Neurol.* 2009;40(3):197–204.
31. Licht DJ, Shera DM, Clancy RR, et al. Brain maturation is delayed in infants with complex congenital heart defects. *J Thorac Cardiovasc Surg.* 2009;137(3):529–536.
32. Ortinau C, Beca J, Lambeth J, et al. Regional alterations in cerebral growth exist preoperatively in infants with congenital heart disease. *J Thorac Cardiovasc Surg.* 2012;143(6):1264–1270.
33. Owen M, Shevell M, Donofrio M, et al. Brain volume and neurobehavior in newborns with complex congenital heart defects. *J Pediatr.* 2014;164(5):1121–1127.e1.
34. Beca J, Gunn JK, Coleman L, et al. New white matter brain injury after infant heart surgery is associated with diagnostic group and the use of circulatory arrest. *Circulation.* 2013;127(9):971–979.
35. Berg C, Gembruch O, Gembruch U, Geipel A. Doppler indices of the middle cerebral artery in fetuses with cardiac defects theoretically associated with impaired cerebral oxygen delivery in utero: is there a brain-sparing effect? *Ultrasound Obstet Gynecol.* 2009;34(6):666–672.
36. Williams IA, Tarullo AR, Grieve PG, et al. Fetal cerebrovascular resistance and neonatal EEG predict 18-month neurodevelopmental outcome in infants with congenital heart disease. *Ultrasound Obstet Gynecol.* 2012;40(3):304–309.
37. Rollins CK, Watson CG, Asaro LA, et al. White matter microstructure and cognition in adolescents with congenital heart disease. *J Pediatr.* 2014. http://dx.doi.org/10.1016/j.jpeds.2014.07.028. pii:S0022–3476(14)00652-0 [Epub ahead of print].
38. Volpe JJ. Encephalopathy of congenital heart disease- destructive and developmental effects intertwined. *J Pediatr.* 2014;164(5):962–965.
39. Kussman BD, Wypij D, Laussen PC, et al. Relationship of intraoperative cerebral oxygen saturation to neurodevelopmental outcome and brain magnetic resonance imaging at 1 year of age in infants undergoing biventricular repair. *Circulation.* 2010;122(3):245–254.
40. Cheng HH, Wypij D, Laussen PC, et al. Cerebral blood flow velocity and neurodevelopmental outcome in infants undergoing surgery for congenital heart disease. *Ann Thorac Surg.* 2014;98(1):125–132.

41. van der Rijken R, Hulstijn W, Hulstijn-Dirkmaat G, Daniëls O, Maassen B. Psychomotor slowness in school-age children with congenital heart disease. *Dev Neuropsychol.* 2011;36(3):388–402.

42. Belsky J, de Haan M. Parenting and children's brain development: the end of the beginning. *J Child Psychol Psych.* 2011;52(4):409–428.

43. Meyers JE, Meyers K. *Rey complex figure test and recognition trial.* Odessa, FL: Psychological Assessment Resources; 1995.

44. Watanabe K, Ogino T, Nakano K, et al. The Rey-Osterrieth complex figure as a measure of executive function in childhood. *Brain Dev.* 2005;27(8):564–569.

45. Marino BS, Lipkin PH, Newburger JW, et al. Neurodevelopmental outcomes in children with congenital heart disease: evaluation and management: a scientific statement from the American Heart Association. *Circulation.* August 28, 2012;126(9):1143–1172.

An Emergent Phenotype: A Critical Review of Neurodevelopmental Outcomes for Complex Congenital Heart Disease Survivors During Infancy, Childhood, and Adolescence

M. *Kharitonova*

Northwestern University Feinberg School of Medicine, Chicago, IL, United States

B.S. *Marino*

Ann & Robert H. Lurie Children's Hospital of Chicago, Chicago, IL, United States

Northwestern University Feinberg School of Medicine, Chicago, IL, United States

OUTLINE

INTRODUCTION

New surgical techniques and advances in cardiopulmonary bypass (CPB), intensive care, and interventional cardiac catheterization have significantly lowered mortality rates in children and adolescents with complex congenital heart disease (CHD).[1,2] Compared to heart-healthy children, complex CHD survivors are at greater risk for neurodevelopmental (ND) deficits that result from both biological and environmental risk factors.[3] Biological risk factors include underlying syndromes or genetic disorders, the circulatory abnormalities specific to the congenital heart defect, the medical and surgical therapies required, and the psychosocial stress of living with a serious chronic disease. Biological risk factors are moderated by environmental risk and resilience factors at home and at school. Developmental concerns among children with CHD may start in infancy but often become more apparent in later childhood and adolescence.[4-9] Overall, complex CHD survivors have a distinctive pattern of ND and behavioral impairments. These are characterized by mild cognitive impairment, deficits in core communication skills and pragmatic language, inattention, hyperactivity and impulsivity, deficits in visual construction and perception, impaired executive functioning, and limitations in both gross and fine motor skills.[10] In addition, there are often accompanying behavioral and emotional issues such as posttraumatic stress symptomatology, anxiety, and depression in both the survivor and the family, which become more apparent with age.[11,12] Many school-age survivors of infant cardiac surgery require supportive services including tutoring, special education, and physical, occupational, and speech therapy. The ND and psychosocial morbidity related to CHD and its treatment often limit ultimate educational achievements, employability, lifelong earnings, insurability, and the quality of life (QoL) for many patients. A significant proportion of patients with complex CHD may need specialized services into adulthood. Incorporation of new stratification methods and clinical evaluation and management algorithms may result in increased surveillance, screening, evaluation, diagnosis, and management of developmental disorder and disability in the complex CHD population and consequent improvement in ND and behavioral outcomes in this high-risk population. With early identification and treatment of developmental disorders and delays, children have the best chance to reach their full potential.[13,14]

The purpose of this chapter is to compare and contrast ND phenotypes in complex CHD survivors across ages and across research centers in North America and Europe. The purpose of this chapter is to compare and contrast ND phenotypes in complex CHD survivors across ages and across research centers in North America and Europe, by focusing on studies published since 1995 and evaluating ND outcomes in CHD survivors. This chapter centers on complex CHD that requires CPB in the neonatal or infant period and examines ND outcomes across the age continuum: infants (birth to 12 months), toddlers (1–3.5 years), preschoolers (3.5–5 years), school-age children (5–12 years), and adolescents (13–19 years). We integrated studies conducted in the North America and Europe. For each age group, refer to Table 5.1 for details about participant demographics and domains assessed in each study.

Following the framework outlined in the American Heart Association 2012 Scientific Statement on Neurodevelopmental Outcomes in Children with Congenital Heart Disease,[3] we describe findings organized by ND domain within each age group, while collapsing geographic regions and types of CHD. These domains included cognitive, motor, neuropsychological, and behavioral–emotional skills, as they develop across infancy, childhood, and adolescence. We conclude each section with a summary of existing findings and overview of current gaps of knowledge that should be investigated in follow-up work. For an overview of ND outcomes across domains and age groups, refer to Table 5.2.

INFANTS

Three investigations across North America and Europe examined ND outcomes in infants who had undergone surgery with CPB several months after corrective surgery.[6,9,15,16] To assess ND in infancy, all of these studies included a version of Bayley Scales of Infant Development (BSID),[17–19] which is a well-established, standardized assessment of cognitive and motor development for children aged 1 through 42 months and yields both a Mental Developmental Index (MDI) (proxy for *General Cognition*) and a Psychomotor Developmental Index (PDI) (proxy for *Motor Skills*). The BSID-III test, used in one of these three studies[15] also includes scores for the *Language* domain. All indices have normative scores of 100 ± 15.

General Cognition

General cognition was not shown to be markedly different for those infants with CHD who had survived heart surgery relative to the heart-healthy controls. The Boston Circulatory Arrest Study (BCAS) found that only 6% (8/143) received scores of 80 or below on the MDI.[6] The average

TABLE 5.1 Review of ND Outcome Studies

Study	Country (Continent)	Number of Participants (n) With CHD	Ages of Participants	Type of CHD	Domain Assessed and Measures Used
Infants					
Ohio State, Columbus, OH[15]	United States (North America)	n = 10	3 months	Complex CHD requiring surgical intervention within the first month of life	• *General Cognition* (BSID-III[19]) • *Motor Skills* (BSID-III[19]) • *Language* (BSID-III[19]) • *Memory* (mobile paradigm[62])
Belfast[16]	Northern Ireland (Europe)	n = 70	8 months	A variety of cyanotic and acyanotic CHD	• *General Cognition* (BSID-II[17]) • *Motor Skills* (BSID-II[17])
Boston Circulatory Arrest Study (BCAS), Boston, MA[6,9]	United States (North America)	n = 171	11–13 months	d-TGA	• *General Cognition* (BSID[18]) • *Motor Skills* (BSID[18]) • *Memory* (Fagan Test of Infant Intelligence[20])
Toddlers					
Single Ventricle Reconstruction Study (SVR-I), National Institute of Health[29]	United States (North America): 15 centers	n = 321	14 months	HLHS or other single, morphological right ventricle anomaly with Norwood procedure	• *General Cognition* (BSID-II[17]) • *Motor Skills* (BSID-II[17])
Western Canadian Complex Pediatric Therapies Follow-up Group[26,27]	Canada (North America)	n = 34–88	18–24 months	TAPVC[26], TGA[27]	• *General Cognition* (BSID-II[17]) • *Audiology* (binaural or bilateral sensorineural hearing loss greater than 40 dB at any frequency from 250 to 4000 Hz)

Continued

II. TOWARD A NEURODEVELOPMENTAL PHENOTYPE

TABLE 5.1 Review of ND Outcome Studies—cont'd

Study	Country (Continent)	Number of Participants (n) With CHD	Ages of Participants	Type of CHD	Domain Assessed and Measures Used
Boston Circulatory Arrest Study (BCAS)[9] Boston, MA	United States (North America)	n = 114	30 months	d-TGA	• General Cognition (MCDI[63]) • Motor Skills (MCDI[63]) • Language (MCDI[63]) • Behavioral–Emotional (MCDI[63])
Single Ventricle Reconstruction II Study (SVR-II), National Institute of Health[28]	United States (North America): 15 centers	n = 203	36 months	HLHS or other single, morphological right ventricle anomaly and planned Norwood procedure	• General Cognition (ASQ[64]) • Motor Skills (ASQ[64]) • Language (ASQ[64]) • Behavioral–Emotional (ASQ[64])
Preschoolers					
Belfast[30]	Northern Ireland (Europe)	n = 90	4 years	A variety of cyanotic and acyanotic CHD	• General Intelligence (WPPSI-R[65]) • Motor Skills (NEPSY[66]) • Language (NEPSY[66]) • Memory (NEPSY[66]) • Processing Speed (WPPSI-R[65]) • Attention (NEPSY[66]) • Visuospatial integration (NEPSY[66]) • Behavioral–Emotional (CBCL[50])
The Children's Hospital of Philadelphia, Philadelphia, PA[31]	United States (North America)	n = 365	4–5 years	Fontan procedure for HLHS and other single ventricle defects (n = 112) and biventricular repair (n = 253)	• General Cognition (WPPSI-II[67]) • Academic achievement (Woodcock-Johnson II[68]) • Visual–Motor Integration (WPPSI-III[67]) • ADHD Symptoms (Inattention Scales of the ADHD-IV Preschool Version[69]) • Behavioral–Emotional (Preschool and Kindergarten Behavior Rating Scales Social Skills Total Score[70]; CBCL[71])

Boston Circulatory Arrest Study (BCAS), Boston, MA[7,32,33]	United States (North America)	$n = 158$[7] $n = 30$[32] $n = 76$[33]	4 years	d-TGA	• *General Cognition* (WPPSI–R[65]) • *Motor Skills* (Peabody Developmental Motor Scales[72]; Grooved Pegboard[73]) • *Language/Speech* (Test for Auditory Comprehension of Language[74]; Receptive One-Word Picture Vocabulary Test[75]; Expressive One-Word Picture Vocabulary Test[76]; Grammatic Closure subtest of the Illinois Test of Psycholinguistic Abilities[77]; play with a toy house to elicit a sample of parent–child conversational talk[32]; produce narratives of personal experience[33]; Index of Productive Syntax[78]; Oral and Speech Motor control Test[79] Mayo Tests for Apraxia of Speech and Oral Apraxia-Children's Battery (selected items)[80]; Goldman–Fristoe Test of Articulation[81]) • *Audiology* (conditioned play audiometry or sound field audiometry)[7]
School-age children					
Boston Circulatory Arrest Study (BCAS), Boston, MA[8]	United States (North America)	$n = 155$	8 years	d-TGA	• *General Cognition* (WISC-III[82]) • *Motor skills* (Grooved Pegboard[83]) • *Language* (Formulated Sentences subtest of the Clinical Evaluation of Language Fundamentals—Third Edition[84], Controlled Oral Word Association Test[85], Verbal Fluency subtest of the McCarthy Scales[86])

Continued

TABLE 5.1   Review of ND Outcome Studies—cont'd

Study	Country (Continent)	Number of Participants (n) With CHD	Ages of Participants	Type of CHD	Domain Assessed and Measures Used
					• *Speech* (Mayo Test for Apraxia of Speech and Oral Apraxia-Children's Battery (selected items)[80]; Oral and Speech Motor Control Test[79]; Goldman–Fristoe Test of Articulation[81]; Auditory Closure subtest of the Illinois Test of Psycholinguistic Ability[77]; Test of Auditory Analysis[87]) • *Memory* (Wide Range Assessment of Memory and Learning[88]) • *Executive Function* (WCST[89]; Trail-Making Test-Intermediate Version[73]) • *Attention* (Test of Variables of Attention, version 6.0.8[90]) • *Visual–Motor Integration* (Developmental Test of VMI, 3rd Revision[91]; Rey–Osterrieth Complex Figure[92]) • *Academic Achievement* (WIAT[93]; Adaptive Functioning scales of the Teacher Report Form)[94] • *Behavioral–Emotional* (CBCL/4-18[50])

Children's Hospital of Philadelphia, PA[38,44]	United States (North America)	n = 35[38] n = 109[44]	5–18 years[38] 5–10 years[44]	A variety of cyanotic and acyanotic CHD	• *General Cognition* (WASI[95]) • *Motor skills* (Grooved Pegboard[83]) • *Memory* (BSRT[96]; BVRT[97]) • *Attention* (Children's Memory Scale[98]; TOL-DX[99]) • *Executive function* (Children's Memory Scale[98]; TOL-DX[99]) • *Visual–Motor Integration* (VMI[91]) • *Behavioral–Emotional* (BASC[100]) • *ADHD Symptoms* (ADHD-IV[101]; BASC)[100]
Belfast[39]	Northern Ireland (Europe)	n = 31	7–8 years	A variety of cyanotic and acyanotic CHD	• *Behavioral–Emotional* (CBCL[45])
Aachen[40–43]	Germany (Europe)	n = 40[42]	7 years	TOF and VSD	• *General Cognition* (K-ABC[102]) • *Motor Skills* (Kiphard and Schilling Body Coordination Test[47]) • *Languge/Speech* (K-ABC[102]; Oral and Speech Motor Control Protocol[79]; Mayo Test of Speech and Oral Apraxia (Children's Battery)[80]; Auditory Closure (AC) subtest of the ITPA[77]; TAAS[103] • *Attention* (ANT[46]) • *Executive Function* (ANT[46]) • *Behavioral–Emotional* (CBCL[50]) • *Audiology* (tone threshold audiometry)

Continued

II. TOWARD A NEURODEVELOPMENTAL PHENOTYPE

TABLE 5.1 Review of ND Outcome Studies—cont'd

Study	Country (Continent)	Number of Participants (n) With CHD	Ages of Participants	Type of CHD	Domain Assessed and Measures Used
Adolescents					
Boston Circulatory Arrest Study (BCAS)[5]	United States (North America)	n = 139	16 years old	d-TGA	• *Academic achievement* (WIAT-II[104]) • *Memory* (General Memory Index of the Children's Memory Scale[98]) • *Attention* (Conners' ADHD scale[105]) • *Executive Function* (Delis–Kaplan Executive Function System[106]; BRIEF[107]) • *Visual–Spatial Skills* (7 subscales of the Test of Visual–Perceptual Skills[108,109]; Developmental Scoring System[110]; Sense of Direction Scale[111]) • *Behavioral–Emotional* (the Reading the Mind in the Eyes Test-Revised[112])

ADHD-IV, Attention Deficit/Hyperactivity Disorder Rating Scale-IV; *ANT*, Attentional Network Test; *ASQ*, Ages and Stages Questionnaire; *BASC*, Behavior Assessment System for Children; *BRIF*, Behavior Rating Inventory of Executive Function; *BSID*, Bayley Scales of Infant Development; *BSRT*, Buschke Selective Reminding Test; *BVRT*, Benton's Visual Retention Test; *CBCL*, Child Behavior Checklist; *CHD*, congenital heart disease; *d-TGA*, dextro-transposition of the great arteries; *HLHS*, hypoplastic left heart syndrome; *ITPA*, Illinois Test of Psycholinguistic Abilities; *K-ABC*, Kaufman Assessment Battery for Children; *MCDI*, Minnesota Child Development Inventory; *NEPSY*, A Developmental NEuroPSYchological Assessment; *TAAS*, Test of Auditory Analysis Skills; *TAPVC*, total anomalous pulmonary venous connection; *TOF*, tetralogy of Fallot; *TOL-DX*, Tower of London, Drexel Version; *VMI*, Visual–Motor Integration; *VSD*, ventricular septal defect; *WASI*, Wechsler Abbreviated Scale of Intelligence; *WCST*, Wisconsin Card Sorting Test; *WIAT*, Wechsler Individual Achievement Test; *WISC*, Wechsler Intelligence Scale for Children; *WPPSI-R*, Wechsler Preschool and Primary Scale of Intelligence-Revised.

TABLE 5.2 Summary of ND Outcomes by Age and Domain

Age Group	Domains Assessed	Outcome	Missing Domain(s)
Infants (birth to 12 months)	General cognition	Scores in the normal range	Executive function
	Motor skills	Deficits observed	Audiology
	Language	Scores in the low-normal range	
	Memory	Deficits observed	
Toddlers (1–3.5 years)	General cognition	Deficits observed	Executive function
	Motor skills	Deficits observed	Memory
	Language	Deficits observed	
	Behavioral–emotional	Deficits observed	
	Audiology	Scores in the normal range	
Preschoolers (3.5–5 years)	General cognition	Scores in the low-normal range	Executive function
	Academic achievement	Scores in the normal range	Memory
	Motor	Deficits observed	
	Language	Some deficits observed for more complex measures	
	Processing speed	Deficits observed	
	ADHD symptoms	Deficits observed	
	Behavioral–emotional	Deficits observed	
	Audiology	Slight deficits observed	

Continued

TABLE 5.2 Summary of ND Outcomes by Age and Domain—cont'd

Age Group	Domains Assessed	Outcome	Missing Domain(s)
School-age children (5–12 years)	General cognition	Scores in the low-normal range	Some domains in single studies only, difficult to generalize
	Academic achievement	Deficits observed	Quality of life
	Language	Scores in the low-normal range	
	Speech	Deficits observed	
	Motor skills	Deficits observed	
	Memory	Deficits observed	
	Executive function	Deficits observed	
	Attention (sustained)	Deficits observed	
	ADHD symptoms	Deficits observed	
	Visual–motor/visual–spatial	Deficits observed	
	Behavioral–emotional	Deficits observed	
	Audiology	Scores in the normal range	
Adolescents (13–18 years)	Academic achievement	Deficits observed	Motor skills
	Executive function	Deficits observed	Audiology
	Memory	Deficits observed	Quality of life
	Visual–spatial skills	Deficits observed	
	Attention (ADHD symptoms)	Deficits observed	
	Behavioral–emotional	Deficits observed	

MDI scores in the BCAS cohort was 105.1 ± 15.0,[6,9] providing no evidence for cognitive deficits early in development. Consistent with this finding, a study in Belfast, Ireland found MDI scores in the normal range (92.9 ± 12.1).[16] Finally, an Ohio State study, which examined the youngest infants at only 3 months of age, found MDI scores for the majority of patients to be in the low-normal range as well (MDI: 88.5 ± 18.1).[3,15]

Motor Skills

In contrast to measures of General Cognition, all but one investigation reported deficits in the motor domain.[6,9,15,16] In BCAS group 20% (28/142) of the children received scores of 80 or below on the PDI.[6,17–19] In addition, a much larger percentage of the BCAS cohort ["neat pincer" grasp (31%) and playing "pat-a-cake" (45%)] failed to reach fine and gross motor milestones compared to the typically developing (TD) population.[9,15] Consistent with this profile, the Belfast study found decreased PDI (75.8 ± 16.9) scores, relative to the TD population.[6,16] In contrast, the Ohio State study reported PDI scores in the low-normal range (PDI: 90.1 ± 7.4).[6,9,15]

Language

In the Ohio State study, which included the youngest infant participants of the infant studies[15,16] language skills were not different from the TD sample. This is not surprising, given that language development is still in its very early stages in early infancy.

Memory

Two studies examined memory performance of infants with CHD. The Fagan Test of Infant Intelligence assesses visual recognition memory (by noting infants' preference to visually fixate on novel over familiar pictures) and predicts subsequent cognitive delay. Failing this test at 3–7 months of age identifies cognitive delays at age 3 years with a moderate level of sensitivity (75%) and specificity (91%).[20] In the BCAS study, 21% (23/107) of infants with CHD failed the Fagan Test.[6] Using a well-studied paradigm of mobile movement by kicking, the Ohio State study examined whether there were differences in how well short-term memories were formed and retained by infants with CHD relative to the TD population. Both the TD and the CHD infants learned at comparable rates (71.4% vs 70% of infants in each group, respectively) to elicit mobile movement by kicking. In contrast, 80% of TD infants retained this knowledge till day 2 of the test, relative to only 43% of infants with CHD, showing evidence of TD group memory for mobile movement, while the CHD group failed to move the mobile above baseline rates.[15]

Summary and Gaps in Knowledge

The above studies suggest a mild deficit in cognition even in infancy, with the more pronounced deficit in the motor domain. Two studies that tested the effects of memory found deficits in this important domain. Studies examining more focused ND domains in infancy, such as executive functioning (measures exist for infants as young as 6–10 months), must be conducted to develop a more precise understanding of the ND profile in infancy.[21–24] In addition, studies of audiologic development are lacking. Given that there is evidence of hearing dysfunction in CHD survivors,[25] understanding when and how these deficits develop starting with the youngest ages is critical.

TODDLERS

Four studies in North America examined ND profiles in toddlers with CHD after cardiothoracic surgery with CPB during infancy. Toddlers are defined as being between 1 and 3.5 years of age (see Table 5.1).[9,26–29]

General Cognition

By toddlerhood, ND deficits no longer appear to be motor domain specific. MDI scores for toddlers with CHD after cardiothoracic surgery with CPB are quite variable, with mean scores below the normal range, and very consistent across studies and centers. The Single Ventricle Reconstruction (SVR-I) Trial found mean MDI score of 89 ± 18 in 14-month-olds.[29] Very similarly, CHD infants in the Western Canadian Complex Pediatric Therapies Follow-up Study (Western Canada) had MDI scores of 89 ± 17 at 2.5 years of age (range: 49–118).[27] In the Western Canada interprovincial cohort, 12–17% of toddlers with CHD showed significant delays (scores <70) on the MDI scale of the BSID-II,[26,27] which is 7.5 times more frequent than would be expected in the TD population. Similarly, scores on the Ages and Stages Questionnaire (ASQ) Problem-Solving domain (used here as a proxy for general cognition) showed significant decreases relative to the TD population (45.4 ± 16.3 vs 55.0 ± 8.2).[28]

Motor Skills

Toddlers with CHD who have undergone cardiothoracic surgery with CPB continue to have moderate deficits in the motor domain during toddlerhood, although in some studies these deficits appear to be less pronounced than in infancy. In the SVR-I study in 14-month-olds, the mean PDI score on the BSID-II[17] was 74 ± 19.[29] In contrast, by 2.5 years of age in the Western Canada study, the scores were more normalized, with a mean PDI score of 92 ± 15 (range: 49–125).[27] Six percent of toddlers with CHD who had undergone cardiothoracic surgery with CPB showed a significant delay (scores <70 (Although the authors considered a delay a score that is 2 SD below the mean (ie, scores <70), typically scores 1 SD below the mean are considered to be "at risk." Thus, a larger proportion of participants can be considered "at risk" using this more liberal but more typical criterion.)) on the PDI of BSID-II,[26,27] a proportion that is 2.7 times more frequent than is expected in the TD populations. On the ASQ Gross and Fine Motor scales, toddlers with CHD showed significant deficits relative to the TD population as well (43.7±16.9 for CHD Gross Motor vs 54.7±9.5 for TD Gross Motor; 38.9±18.5 for CHD Fine Motor vs 52.5±10.9 for TD Fine Motor).[28]

Language

Children with CHD who had undergone cardiothoracic surgery with CPB showed 2–4 months delay in their expressive language development.[9] Overall, moderate delay in expressive language was noted, such that 6% of children did not produce any two-word utterances (relative to 0% in the TD population).[9] The CHD survivors also showed deficits on the Communication scale of the ASQ (47.0±14.7 for CHD vs 54.3±7.8 for the TD group).[28]

Behavioral–Emotional Development

Children with CHD who had undergone cardiothoracic surgery with CPB did not have more behavioral difficulties than TD children, with the exception of those who were not yet combining words at 2.5 years of age.[9] At 3 years of age, 8–22% of children in the SVR-II cohort were found to have behaviors in the at-risk or clinically abnormal range, when measured by the BASC-2.[28] The same study found that toddlers with CHD who had undergone cardiothoracic surgery with CPB have scores on the Personal–Social subscale of the ASQ that is below that of the TD sample (47.0±13.7 for CHD vs 53.5±7.4 for the TD group).[28]

Audiology

There was no evidence of hearing impairment in children with CHD who had undergone surgery with CPB, defined as binaural or bilateral sensorineural hearing loss greater than 40 dB at any frequency from 250 to 4000 Hz among CHD participants.[26,27]

Summary and Gaps in Knowledge

Toddlers with CHD show moderate deficits across a range of domains, no longer limited to motor skills. Specifically, a delay in at least one domain [scores less than 2 standard deviations (SD) below the normative mean] were detected in 51% of the SVR-1 cohort at 3 years of age on the

ASQ; 20% on the Communication Scale, 30% on the Gross Motor Scale, 35% on the Fine Motor Scale, 24% on the Problem Solving Scale, and 17% on the Personal–Social Scale.[28] Only audiologic function appeared to be spared. Studies focusing on executive function and memory abilities in toddlers are needed.

PRESCHOOLERS

Three studies (one in Belfast, Ireland and two in the United States) examined ND profiles in preschoolers with CHD who had undergone cardiothoracic surgery with CPB during infancy. Preschoolers are defined as being 4–5 years of age (see Table 5.1).[7,30–33]

General Cognition

In the Belfast study, preschoolers with CHD who had undergone cardiothoracic surgery with CPB scored in the normal range on measures of general intelligence, attention, memory, cognitive speed, and number skills.[30] In a study performed at the Children's Hospital of Philadelphia (CHOP), preschoolers with CHD scored in the low-normal range on all IQ measures, with a larger than expected variability across all components (eg, full scale IQ: 95.3 ± 19.0, with the normative scores being 100 ± 15.0). These preschoolers scored in the normal range for measures of school readiness (96.0 ± 20.6 for mathematics and 106.0 ± 16.8 verbal), although increased variability was noted.[31] Similarly, in the BCAS, 4-year-old CHD survivors had IQ scores in the low-normal range (92.6 ± 14.7, 95.1 ± 15.0, and 91.6 ± 14.5 for full-scale, verbal, and performance components, respectively) and were significantly lower than the normative mean of 100.[7]

Academic Achievement

In the CHOP study academic achievement scores were in the normal range, with variability that was larger than that expected based on normative data. As such, mathematics achievement scores were 96.0 ± 20.6 and reading achievement scores were 106.0 ± 16.8.[31] This finding supports other data suggesting relative sparing of mathematics and reading abilities in survivors of complex CHD.

Motor Skills and Sensorimotor Integration

In the Belfast study, children with the most complex CHD were at the greatest risk (0.75 SD below the normative mean) for sensorimotor problems.[30] In the CHOP study, preschoolers scored in the low-normal range on a measure of visual motor integration, with a larger than expected variability (scores: 92.7 ± 18.2, with the normative scores being 100 ± 15.0).[31] Consistent with this finding, the BCAS preschoolers also found lowest scores on the visual–spatial and the visual–motor integration skills subtests. Moreover, preschooler CHD surgical survivors scored only in the 9th and 4th percentile on measures of gross and fine motor skills, respectively, consistent with the idea of a moderate motor deficit in early childhood[7] in both the motor skills themselves and the ability to integrate motor and sensory information.

Language

In the Belfast study, preschoolers with the most complex CHD who had undergone cardiothoracic surgery with CPB showed a slight deficit (half of SD below the normative mean) on measures of language when measured by the Developmental NEuroPSYchological (NEPSY) Assessment, but not other measures of verbal reasoning.[30] The BCAS also reported scores in the normal range for the Wechsler Preschool and Primary Scale of Intelligence (WPPSI) subtests of Comprehension and Sentences in preschool survivors. In addition, although both Verbal and Performance [visual–spatial (nonverbal)] IQ scores were lower for the CHD surgical group than for the TD sample, Verbal IQ scores were significantly higher than Performance IQ scores.[7] In a similar vein, when BCAS preschoolers were compared with their heart-healthy TD peers on basic language measures and on proportions of symbolic and nonsymbolic talk, the CHD group was found to have more concrete and less symbolic speech patterns.[32] Finally, when asked to provide biographical narratives, BCAS preschoolers produced fewer narrative components than TD children.[33] These patterns suggest difficulties integrating more complex contextual information in a coherent whole in a goal-directed manner; a difficulty that is foreshadowing the difficulties in executive function that are observed at older ages.

Processing Speed

In the Belfast study, complex CHD patients scored in the low-normal range on measures of processing speed. Specifically, preschool CHD surgical survivors scored between 8.7 ± 2.7 and 9.6 ± 3.1 points on measures of processing speed and between 8.1 ± 3.3 and 9.6 ± 2.4 on measures of nonverbal reasoning.[30]

ADHD Symptomatology

In the CHOP study preschool CHD surgical survivors had more symptoms of both inattention (6.3 ± 5.5) and impulsivity (7.2 ± 5.6) than the normative sample (normative scores: 5.24 ± 6.02 and 5.67 ± 6.51, respectively),[31] suggesting a larger propensity to developing attention deficit hyperactivity disorder (ADHD).

Behavioral–Emotional Development

In the Belfast study, children with the most complex CHD had the highest scores on Child Behavior Checklist (CBCL) Total Problem behavior score, thus reflective of more behavioral problems in the CHD survivors; these scores were marginally higher than those of the heart-healthy sample.[30]

Audiology

Abnormal hearing, defined as a bilateral increase in threshold of greater than or equal to 16 dB for frequencies of 1–4 kHz, was observed in 8–12% of the BCAS sample. Abnormal hearing was associated with deficits of 8 points on Verbal and Performance IQ scales.[7]

Summary and Gaps in Knowledge

By the time children enter preschool (ages 4–5 years), the experience of having complex CHD with subsequent surgical treatments modifies children's development in several major domains. While their general IQ scores were in the low-normal range, there is a striking pattern of increased variability in scores in the CHD surgical survivors relative to heart-healthy peers. Deficits were most pronounced in domains requiring integration, such as visual–spatial processing, or developing complex narratives or symbolic play. Child CHD surgical survivors also showed greater propensity toward ADHD-like symptoms. This, together with their difficulty in integrating information in a goal-directed manner foreshadows difficulties in executive function that are observed at later ages (see subsequent sections). More focused measures of executive function should be included in follow-up work to more directly test when difficulties with executive function first emerge.[34-37]

SCHOOL-AGE CHILDREN

Four studies (one in Belfast, Ireland, one in Aachen, Germany, and two in the United States) examined ND profiles of school-age children with CHD who had undergone cardiothoracic surgery with CPB during infancy. School-age children are defined as being 5–12 years of age, with most studies focusing on 7–8-year-olds (see Table 5.1)[8,38-43].

General Cognition

In the BCAS sample the mean IQ was in the low-normal range [WISC-III Full-Scale IQ scores of 97.1 ± 15.3 (range 62–138)]. Performance (nonverbal) IQ scores (94.9 ± 14.3) were significantly lower than Verbal IQ score

(99.8 ± 16.6); only the former were significantly lower than the expected population normative data.[8] In contrast, in the CHOP study, Wechsler Abbreviated Scale of Intelligence (WASI) IQ scores were comparable to that of the population mean of 100 (mean scores ranging between 104 and 109), depending on the type of CHD.[38] In the Aachen study, CHD participants scored in the low-normal range on the IQ measures (92.2 ± 12.7), with 22.5% of participants being in the "at risk" category; scoring less than 1 SD (15) below the normative mean of 100. Given the normal distribution of population IQ scores, 16% of children are expected to score lower than 1 SD below the mean. Thus, while the majority of CHD survivors at school age have general intelligence within the normal range, the distribution of IQ scores is shifted to the left, resulting in a larger proportion of the CHD survivors being "at risk" or having low IQ compared to the general population.

Academic Achievement

Across studies, CHD survivors have lower academic achievement than heart-healthy controls as measured by standardized tests. In the BCAS sample, both the Reading Composite ($P < 0.001$) and Mathematics Composite ($P < 0.02$) scales were significantly lower than population norms. Consistent with this finding, 37% of the BCAS children were receiving remedial services in school and 10% had repeated a grade.[8] In the CHOP study, 49% of participants were receiving remedial academic services and 15% were in a special-education classroom.[44] Similarly, in the Belfast study, academic performance (measured by the CBCL teacher-completed competence form[45]) showed significantly impaired performance by CHD survivors relative to heart-healthy siblings.[39] Finally, in the Aachen studies, 30% of CHD survivors showed a decrement in academic performance. Thus, similarly to measures of general intelligence, at least a subset of CHD survivors are showing significant impairments in their academic skills and learning.

Executive Function

Executive functioning is particularly impaired in school-age CHD survivors. In the BCAS sample, CHD survivors were impaired on a measure of cognitive flexibility (Wisconsin Card Sorting Task), such that a large proportion of children (between 25% and 46% of the sample, depending on type of dependent measure) was in the "at risk" category for executive function deficits (ie, scored at least 1 SD below the general population mean). These numbers are higher when compared to 16% of children expected to be "at risk" in the general population.[8] Similarly, in the CHOP study, CHD participants scored significantly lower on the Tower of London (TOL-DX) test (scores ranging from 86 to 94, with the population mean of 100).[38] Executive functions were measured in the Aachen sample with the Attentional

Network Test (ANT[46]), which provides three distinct metrics of attention: alerting, orienting, and conflict. Only the latter component showed an impairment for participants with CHD relative to heart-healthy controls.[40] The conflict metric and the associated neural network are most closely linked with the *executive function* systems, as they rely on the need to ignore conflicting information in the service of task-relevant goals.[46]

Attention

Sustained attention was difficult for the CHD children in the BCAS, with rates of errors of omission more than double those of the norming samples (ie, CHD children failed to respond in a sustained manner significantly more frequently than heart-healthy controls), and the rates of errors of commission were more than double, ie, CHD children made more than twice as many mistakes; finally, response times were about 1 SD slower for the CHD sample, relative to the population mean.[8] Other studies (CHOP and Aachen) utilized measures of attention that test *controlled* or *executive* components of attention (relative to the more "pure" *sustained attention* metrics utilized in BCAS). Results from these measures are reported in the Executive Function section above, as these measures are strongly correlated with more standard measures of executive function (see Discussion below for more on how psychological constructs and measures are interrelated).

Memory

There are less data (and less consensus on the existing data) regarding memory abilities in CHD survivors, with only some of the studies reporting memory deficits. In the BCAS, the mean Memory Screening Index score (90.0 ± 15.3) was significantly lower than the expected population mean of 100 ($P < 0.001$). The same was true for mean scores on all subtests with the exception of Verbal Memory, consistent with the findings of Verbal IQ being spared in this population.[8] In contrast, memory was not found to be impaired in the CHOP study[38]; however, we should note that this study used memory measures that were developed in the 1970s; memory measurement science has improved tremendously in its sensitivity, and therefore, a more contemporary measure might have observed finer-tuned differences that were not observed in the CHOP study. Memory was not explicitly assessed in the Belfast and Aachen studies.

Motor Skills and Visual–Motor Integration

CHD survivors continue having difficulties with motor skills as they reach school age. In the Aachen study, only 50% of the CHD survivors scored in the normal range on the Kiphard and Schilling Body Coordination Test,[47] with 42.5% of participants being at least "at risk" category

(1 SD below the expected mean).[42] In the BCAS, similar to other domains, scores on the visual–motor integration measures (Developmental Test of Visual–Motor Integration; VMI[48]) were in the low-normal range, with scores ranging from 89.0 ± 10.6 to 91.4 ± 11.6 (depending on type of CHD), with the overall score for the CHD group being only in the 25th percentile for the general population.[8] In the CHOP sample, using the same VMI test, school-aged CHD survivors also scored in the low-normal range (95 ± 13, relative to 100 ± 15 for the general population).

Language

Language function is relatively well preserved in the CHD sample. On measures of expressive and receptive language in the Aachen studies, CHD participants again scored in the low-normal range (92.9 ± 17.7 for expressive, and 92.6 ± 15.1 for receptive language measures), with 25% of participants being in the "at risk" category for both expressive and receptive language difficulties.[42] In the BCAS, some of the language subtests (eg, Formulated Sentences or Verbal Fluency) showed a decrease in the overall performance relative to heart-healthy controls.[8] These more complex measures require integration among skill sets in a goal-directed manner, and thus, also require executive function, which are often decreased in the CHD survivor population.

Speech

In the Aachen studies, there were observed deficiencies in oral and speech motor control problems. These impairments seemed to improve with age, such that normal oral and speech motor control function were observed in 29% of all children with a mean age of 6.5 years, in contrast to 43% of children with normal performance and a mean age of 8.3 years. However, due to the cross-sectional nature of the sample, a developmental effect cannot be established. Despite relatively spared speech function, between 25% and 50% of children receive speech therapy and the prevalence of speech-related developmental risks and disorders exceeded the 2–10% that is estimated to occur in heart-healthy participants.[43] Similarly, in the BCAS cohort 34% of children were reported by their parents to have seen a speech pathologist. Speech issues were found to be more prevalent in those children with more severe CHD who had a great total circulatory arrest time.[8]

ADHD Symptomatology

School-age survivors of complex CHD are at an increased risk for having symptoms of ADHD. In one CHOP study,[44] 30% of the parents reported high-risk scores for inattention, while 29% reported high-risk scores for hyperactivity. Moreover, the number of children with clinically significant scores for inattention and hyperactivity was 3–4 times higher than that observed in the normative population. Another CHOP study found 69.5% of hypoplastic left heart syndrome (HLHS) children met criteria for ADHD based on neurological examinations, which is substantially higher than the 3–5% incidence rate of ADHD in the general population.[49] The Aachen study assessed inattention and hyperactivity using the Externalizing problems and Attention problems subscales of the CBCL.[50] CHD survivors scored significantly higher on both of these measures relative to heart-healthy controls.[40] Moreover, the number of parent-reported externalizing and attention problems on CBCL was inversely correlated with performance levels on the conflict scale of the ANT, which is considered to be a measure of executive function.

Behavioral–Emotional Development

As in earlier ages, there were moderately more behavioral problems observed in children with CHD, relative to heart-healthy controls. For example, in the CHOP study, CHD participants' average scores on the BASC measure were 43 and 46 (depending on type of CHD), with the population mean of 50 (lower scores reflect more problems on this measure). In the Aachen study, CHD participants showed significantly more internalizing (anxiety and depression), externalizing (aggression), and total behavioral problems (social and attention problems) and reduced results with respect to school performance and total competence compared to the normative population.[41] Similarly, in the Belfast study, CHD survivors had more total behavioral problems (with 27% of CHD survivors being in the clinically significant range for behavior problems) relative to their heart-healthy siblings (none of whom met criteria for clinical levels).[39]

Audiology

Audiology scores were normal in the Aachen studies, such that at the time of assessment, all children had normal faculty of hearing as evaluated by tone threshold audiometry.[42] There are little data about hearing impairment in the CHD population and more work in this area is required.

Summary and Gaps in Knowledge

As children age, the deficits observed across a range of domains become more apparent. Language and some of the general intelligence measures elicit relatively spared performance scores. In contrast, measures that rely on integration of information in goal-directed fashion, such as executive function, speech, visual–motor integration, and social–emotional or behavioral development show significant deficits in survivors of CHD. As discussed in more detail in the General Discussion section, there are several possibilities for why deficits appear to be more severe as children age. It is possible that early insult to the brain (eg, due to hypoxia and ischemia) mostly impairs networks that are important for functional information integration, which might be less necessary for successful function in infancy and very young childhood. It is also possible that the measures used to assess ND profiles in older children are more sensitive, reliable, and valid than those used in earlier childhood, and thus an appearance of more severe deficits appears due to measure sensitivity. Finally, we should keep in mind that in many cases often a single study addresses each domain, and therefore, generalizing across samples or across geographic locales is proving difficult. Thus, future research should investigate more ND domains per study, ideally with comparable measures across studies and regions, to enable a more precise understanding of the severity of ND impairments in CHD survivors in school age.

ADOLESCENTS

There was only one study that globally assessed adolescent ND and psychosocial functioning across a broad set of domains. The BCAS investigated the ND profile of the cohort with the dextro-transposition of the great arteries during adolescence at age 16 years (see Table 5.1).[5]

Academic Achievement

Similar to earlier ages, academic achievement scores were in the low-normal range (scores in the 90s), but significantly below the normative mean of 100, for both reading and mathematics components. Twenty-seven percent of CHD participants scored in "at-risk" range (<1 SD below the normative mean) for reading and mathematics, respectively.[5]

Executive Function

Consistent with the other domains, the CHD executive function scores (9.0 ± 2.1) were significantly below the expected normative mean of 10. Moreover, there was a large proportion of participants (13–38%, depending on whether measurement was by self-report, parent based, or teacher based) who met criteria for "clinical concern" on the BRIEF measure.[5]

Memory

Memory scores (90.4 ± 18.5) were also significantly below the normative mean of 100. Thirty-five percent of CHD participants were categorized "at risk" with scores <1 SD below the expected normative mean.

Visual–Spatial Skills

The mean visual–spatial skills score for the CHD group (85.6 ± 16.5) was 1 SD below the normative mean (ie, in the "at risk" category), with 54% of participants scoring even further below the score of 85 ("at risk"). However, on two other measures of visual–spatial skills, CHD adolescents did not differ from controls; therefore, sensitivity of a test must be considered before conclusions can be drawn.

Attention (ADHD Symptoms)

The CHD adolescents had attention scores (53.6 ± 13.0) that were significantly higher than the normative mean of 50, with 19% of the CHD adolescents having an attention score above the cutoff for clinical concern.

Behavioral–Emotional Development

On the Reading the Mind in the Eyes test, the mean score in the CHD treatment group was lower than that of the control group ($P = 0.03$). The scores on the Adult Autism Spectrum Quotient questionnaire were significantly higher in the CHD treatment group, suggesting that early neurological insult has far-reaching behavioral consequences.

Summary and Gaps in Knowledge

The one existing study of ND outcomes in 16–year-olds paints a concerning picture. The deficits observed in school-age children on a number of domains requiring integration of information (eg, executive function, visual–motor integration, behavioral–emotional development, and speech) appear to persist and/or worsen into adolescence, which could have far-reaching effects on school performance and future vocational or higher-educational opportunities and ultimately self-sufficiency as an adult. It is important to note that these adolescent data were generated at a single institution and obtained from a single CHD group (d-transposition of the great arteries). Thus, future research must focus on how generalizable these data from the BCAS are relative to the broader population of adolescent CHD survivors. If the BCAS adolescent data are supported by other studies on CHD adolescent survivors, these data have significant implications on transition of care, self-sufficiency, and life-success for young adults with complex CHD.

GENERAL DISCUSSION

Summary and Future Directions

Across a number of studies from both North America and Europe, survivors of complex CHD show a range of ND deficits across a wide range of cognitive domains (see Table 5.2). These include but are not limited to general cognitive deficits, speech, visual–motor integration, and difficulties with tasks requiring executive function. Survivors of complex CHD are also at an increased risk of meeting criteria for ADHD, as well as other behavioral–emotional issues (ie, externalizing and internalizing disorders) in later childhood and adolescence. These findings are consistent across a wide range of geographic locations and study centers. In general, more severe CHD (eg, cyanotic or single ventricle) is accompanied by more severe deficits than less severe CHD (eg, acyanotic), but the general pattern of results holds across a wide range of types of complex CHD.

Literature reviewed in this chapter suggests that deficits intensify across time by becoming more present and pronounced in later childhood and adolescence, when compared to infancy and toddlerhood. It is indeed possible that neurological insults that occurred early in life (prenatally, or as a result of surgical interventions in infancy) do not manifest fully until relatively more complex tasks are needed to be implemented in later childhood. Nevertheless, an alternative possibility must be considered: ND constructs appear to be less differentiated in early childhood[34,36] and tests that are intended to measure such constructs also tend to be less sensitive and less reliable in very early childhood. An additional concern is that there are only a handful of longitudinal studies that examine ND phenotypes of CHD survivors across development.[51] Longitudinal cohort studies examining ND outcomes across childhood and adolescence that rely on the more precise contemporary measurement instruments of neurocognitive function are imperative to more definitely understand this phenotype as CHD survivors grow into adulthood.

A related issue that makes precise understanding of the ND phenotype of CHD survivors difficult is the notion that cognitive domains are not isolated entities, but are rather interrelated constructs that rely on overlapping sets of neural networks.[52–54] Instruments that measure *attention* in one study might be used to assess aspects of *executive function* in another study. Use of "common currency measures" is needed to more definitively identify the ND phenotype as it develops across time. NIH

Toolbox provides one option for such "currency," with instruments that assess Cognition, Emotion, Motor, and Sensation domains in participants ages 3–70 with high level of reliability and validity[34] and the availability of national norms.[55] Combining the use of sophisticated neuroimaging tools with common-currency neurocognitive measures will help shed light on the underlying causes of the ND phenotype seen in CHD surgical survivors.

ND and Psychosocial Impact on Health-Related QoL

Importantly, in addition to these neurocognitive effects, QoL is more significantly affected in pediatric survivors of complex CHD. Specifically, lower full-scale IQ (intelligence) and lower performance in reading and mathematics (academic achievement) were associated with lower parent-reported psychosocial QoL scores at 8 years of age.[56] Moreover, children with Fontan palliation for HLHS displayed not only significant delays in communication and motor skills, but also lower parent-reported psychosocial QoL scores.[57] Of note, both of these studies used a generic QoL instrument to measure psychosocial QoL, which may not be as sensitive or accurate as a cardiac disease-specific instrument.[58] Recently, Marino et al. demonstrated that lower levels of executive functioning, gross motor ability, and presence of internalizing disorders, such as anxiety or depression significantly predicted lower QoL scores after controlling for patient demographics and important clinical covariates.[59] Executive functioning, gross motor ability, and internalizing disorders accounted for 42–50% of the variance in both *patient and parent proxy-reported* QoL Psychosocial Impact subscale scores. In addition, executive dysfunction accounted for 37–54% of the variation noted in patient and parent-reported PedsQL School Functioning QoL subscale score. These factors may be key drivers of QoL in complex CHD surgical survivors and should be targets for future intervention.[58]

A Clinical Agenda to Improve ND and Psychosocial Outcomes

For ND evaluation to be a routine component of the care of children with CHD, it has to be efficient, beneficial, and effective for all involved. Identifying key professionals to take responsibility for, and champion the introduction of, such routine evaluation is a crucial first step.[109] The methods for collecting data need to be tailored to the patients and the clinical setting and should allow respondents to provide information independently of the clinical team. Strategies to maximize completion without alienating patients and families should be put in place, taking account of the cost implications of such strategies, and it is important that data collection is conducted in a robust manner so that the information collected is reliable, valid, clinically useful, and of research quality. Before data collection begins, it is important to develop a protocol for administration of the questionnaires, transfer of the data to an appropriate database, scoring and interpretation of the findings, and a mechanism for disseminating the findings to relevant parties. Variability in the way in which data are collected needs to be minimized so that findings are more likely to be generalizable to other settings and/or populations. Using identical (or at least comparable) neuropsychological paradigms across study centers will enable more reliable comparisons and thus more valid conclusions.

Demonstration of the utility of ND assessment should drive the implementation of routine ND evaluation at clinic visits or through separate cardiac ND programs as part of a structured surveillance and screening program, thus allowing for targeted referral, intervention, and follow-up. If the introduction of ND evaluation is to become an integral part of routine care, it is essential that all of the stages of data collection, analysis, and dissemination are simple, quick, and well managed, ensuring that they are optimally tailored to the institution and clinicians involved, so that the benefits are readily apparent in terms of improved patient outcomes and effective management of resources.

A Research Agenda to Improve ND and Psychosocial Outcomes

Before successful interventions can be developed, precise understanding of the causes of the broad ND deficits observed in CHD survivors need to be established both at the psychological and at the neural levels. Generic neuropsychological measures are not "process pure": successful performance on these tests requires a number of abilities and skills that are well separated in these measures. They are thus not particularly useful at elucidating precise *mechanism(s)* that give rise to observed behaviors. For example, difficulties in executive function and visual–motor integration tasks might be due to completely distinct neurocognitive deficits. In contrast, a common deficit might give rise to both of these (and other) observed impairments. For example, maintaining relevant goal information in working memory explains why executive function and processing speed measures are often correlated.[61] Using relatively more "process pure" measures developed **recently** by cognitive psychologists, **along with** sophisticated neuroimaging tools will help shed light on the precise nature of the underlying reasons behind the broad-ranging ND impairments in CHD survivors.

The goal of ND outcomes research should focus on interventions to improve ND and psychosocial outcomes through prevention and treatment. Specifically, research is required to elucidate the links between ND, psychosocial, and physical morbidity factors to identify functional deficits that may be prevented or mitigated through intervention. It is also crucial to demonstrate that research quality ND assessment in the clinical setting is feasible, meaningful, clinically useful, and can be translated into clinical benefit. In particular, algorithms for stratifying patients into high- and low-risk groups in terms of ND, psychosocial, and physical dysfunction need to be developed so that interventions can be targeted for those at high risk. Ultimately, interventions need to result in clinically meaningful and sustainable changes in ND outcomes. Further areas for research include the development and use of responsive, disease-specific ND tools that can be used to assess the benefit of surgical, medical, and other therapeutic interventions, with such a measure being routinely included in all drug, device, and surgical trials. A wide range of tasks intended to assess functioning of specific cognitive subdomains (such as aspects of executive function, like inhibitory control, working memory, and cognitive flexibility[36,60]) have been developed by cognitive and developmental psychologists and should be utilized to enhance understanding of both the affected aspects of cognition and the efficacy of proposed interventions.

As survival rates continue to improve and the CHD population ages, new disease-specific measures for adults with CHD are also needed to allow continued evaluation of this growing population across the lifespan. There is a broad arsenal of neurocognitive measures developed for adults by cognitive psychologists in the last few decades that enable studying the function of very precise cognitive processes. These measures are more elaborate and fine-tuned, with typically greater psychometric qualities for older adolescents and adults, which enhances their usability for the aging CHD population. Using recently developed sophisticated neuroimaging methods in adults will also shed light on the neural substrates of both the observed deficits and the potential recovery of both structure and function that results from targeted interventions.

References

1. Mahle WT, Spray TL, Wernovsky G, Gaynor JW, Clark 3rd BJ. Survival after reconstructive surgery for hypoplastic left heart syndrome: a 15-year experience from a single institution. *Circulation*. 2000;102:136–141.
2. Jacobs JP, Quintessenza JA, Burke RP. Analysis of regional congenital cardiac surgical outcomes in Florida using the Society of Thoracic Surgeons Congenital Heart Surgery Database. *Cardiol Young*. 2009;19(4):360–369.
3. Marino BS, Lipkin PH, Newburger JW, et al. Neurodevelopmental outcomes in children with congenital heart disease: evaluation and management: a scientific statement from the American heart association. *Circulation*. 2012;126(9):1143–1172. http://dx.doi.org/10.1161/CIR.0b013e318265ee8a.
4. Mussatto KA, Hoffmann RG, Hoffman GM, et al. Risk and prevalence of developmental delay in young children with congenital heart disease. *Pediatrics*. 2014;133(3): e570–e577.

5. Bellinger DC, Wypij D, Rivkin MJ, et al. Adolescents with d-transposition of the great arteries corrected with the arterial switch procedure. *Circulation*. 2011;124: 1361–1369.
6. Bellinger DC, Jonas RA, Rappaport LA, et al. Developmental and neurologic status of children after heart surgery with hypothermic circulatory arrest or low-flow cardiopulmonary bypass. *N Engl J Med*. 1995;332(9):549–555.
7. Bellinger DC, Wypij D, Kuban KCK, et al. Developmental and neurological status of children at 4 years of age after heart surgery with hypothermic circulatory arrest or low-flow cardiopulmonary bypass. *Circulation*. 1999;100:526–532.
8. Bellinger DC, Wypij D, duPlessis AJ, et al. Neurodevelopmental status at eight years in children with dextro-transposition of the great arteries: the Boston Circulatory Arrest Trial. *J Thorac Cardiovasc Surg*. 2003;126(5):1385–1396. http://dx.doi.org/10.1016/S0022-5223(03)00711-6.
9. Bellinger DC, Rappaport LA, Wypij D, Wernovsky G, Newburger JW. Patterns of developmental dysfunction after surgery during infancy to correct transposition of the great arteries. *J Dev Behav Pediatr*. 1997;18(2):75–83.
10. Marino BS. New concepts in predicting, evaluating, and managing neurodevelopmental outcomes in children with congenital heart disease. *Curr Opin Pediatr*. 2013;25(5):574–584.
11. Karsdorp PA, Everaerd W, Kindt M. Psychological and cognitive functioning in children and adolescents with congenital heart disease: a meta-analysis. *J Pediatr Psychol*. 2007;32(5):527–541.
12. Latal B, Helfricht S, Fischer JE, Bauersfeld U. Psychological adjustment and quality of life in children and adolescents following open-heart surgery for congenital heart disease: a systematic review. *BMC Pediatr*. 2009;9(1):6.
13. Guralnick MJ. Effectiveness of early intervention for vulnerable children: a developmental perspective. *Am J Ment Retard*. 1998;102(4):319–345.
14. Guralnick MJ. Prevention and early intervention in the social inclusion of children and young people. *J Appl Res Intellect Disabil*. 2005;18:313–324.
15. Chen CY, Harrison T, Heathcock J. Infants with complex congenital heart diseases show poor short-term memory in the mobile paradigm at 3 months of age. *Infant Behav Dev*. 2015;40:12–19.
16. McCusker CG, Doherty NN, Molloy B, et al. A controlled trial of early interventions to promote maternal adjustment and development in infants born with severe congenital heart disease. *Child Care Health Dev*. 2010;36(1):110–117.
17. Bayley N. *Bayley Scales of Infant Development: Manual*. Psychological Corporation; 1993.
18. Bayley N. *Manual for the Bayley Scales of Infant Development*; 1969.
19. Bayley N. *Bayley Scales of Infant Development and Toddler Development*; 2006.
20. Fagan JF, Singer LT, Montie JE, Shepherd PA. Selective screening device for the early detection of normal or delayed cognitive development in infants at risk for later mental retardation. *Pediatrics*. 1986;78(6):1021–1026.
21. Oakes LM, Baumgartner HA, Barrett FS, Messenger IM, Luck SJ. Developmental changes in visual short-term memory in infancy: evidence from eye-tracking. *Front Psychol*. 2013;4:1–13. http://dx.doi.org/10.3389/fpsyg.2013.00697/abstract.
22. Holmboe K, Pasco Fearon RM, Csibra G, Tucker LA, Johnson MH. Freeze-Frame: a new infant inhibition task and its relation to frontal cortex tasks during infancy and early childhood. *J Exp Child Psychol*. 2008;100(2):89–114. http://dx.doi.org/10.1016/j.jecp.2007.09.004.
23. Diamond A, Doar B. The performance of human infants on a measure of frontal cortex function, the delayed response task. *Dev Psychobiol*. 1989;22(3):271–294.
24. Diamond A. Lifespan Cognition: Mechanisms of Change. In: Ellen Bialystok, Fergus I.M. Craik. *Google Books. Lifespan Cognition: Mechanisms of Change*; 2006.

25. Mann T, Adams K. Sensorineural hearing loss in ECMO survivors. *J Am Acad Audiol.* 1998;9(5):367–370.

26. Alton GY, Robertson CM, Sauve R, et al. Early childhood health, growth, and neurodevelopmental outcomes after complete repair of total anomalous pulmonary venous connection at 6 weeks or younger. *J Thorac Cardiovasc Surg.* 2007;133(4):905–911. e3.

27. Freed DH, Robertson CM, Sauve RS, et al. Intermediate-term outcomes of the arterial switch operation for transposition of great arteries in neonates: alive but well? *J Thorac Cardiovasc Surg.* 2006;132(4):845–852. e2.

28. Goldberg CS, Lu M, Sleeper LA, et al. Factors associated with neurodevelopment for children with single ventricle lesions. *J Pediatr.* 2014;165(3):490–496. http://dx.doi.org/10.1016/j.jpeds.2014.05.019. e8.

29. Newburger JW, Sleeper LA, Bellinger DC, et al. Early developmental outcome in children with hypoplastic left heart syndrome and related anomalies: the single ventricle reconstruction trial. *Circulation.* 2012;125(17):2081–2091.

30. McCusker CG, Doherty NN, Molloy B, et al. Determinants of neuropsychological and behavioural outcomes in early childhood survivors of congenital heart disease. *Arch Dis Child.* 2007;92(2):137–141.

31. Gaynor JW, Ittenbach RF, Gerdes M, et al. Neurodevelopmental outcomes in preschool survivors of the Fontan procedure. *J Thorac Cardiovasc Surg.* 2014;147(4):1276–1283.

32. Ovadia R, Hemphill L, Winner K, Bellinger D. Just pretend: participation in symbolic talk by children with histories of early corrective heart surgery. *Appl Psycholinguistics.* 2000;21(03):321–340.

33. Hemphill L, Uccelli P, Winner K, Chang C-J, Bellinger D. Narrative discourse in young children with histories of early corrective heart surgery. *J Speech Lang Hear Res.* 2002;45(2):318–331.

34. Zelazo PD, Bauer P. *National Institutes of Health Toolbox Cognition Battery (NIH Toolbox CB).* Wiley-Blackwell; 2013.

35. Kharitonova M, Chien S, Colunga E. More than a matter of getting "unstuck": flexible thinkers use more abstract representations than perseverators. *Dev Sci.* 2009;12(4):662–669.

36. Wiebe SA, Sheffield T, Nelson JM, Clark CAC, Chevalier N, Espy KA. The structure of executive function in 3-year-olds. *J Exp Child Psychol.* 2011;108(3):436–452. http://dx.doi.org/10.1016/j.jecp.2010.08.008.

37. Espy KA, Sheffield TD, Wiebe SA, Clark CA, Moehr MJ. Executive control and dimensions of problem behaviors in preschool children. *J Child Psychol Psychiatry.* 2011;52(1):33–46.

38. Quartermain MD, Ittenbach RF, Flynn TB, et al. Neuropsychological status in children after repair of acyanotic congenital heart disease. *Pediatrics.* 2010;126(2):351–359.

39. McCusker CG, Armstrong MP, Mullen M. A sibling-controlled, prospective study of outcomes at home and school in children with severe congenital heart disease. *Cardiol Young.* 2013;23:507–516.

40. Hövels-Gürich HH, Konrad K, Skorzenski D, Herpertz-Dahlmann B, Messmer BJ, Seghaye M-C. Attentional dysfunction in children after corrective cardiac surgery in infancy. *Ann Thorac Surg.* 2007;83(4):1425–1430. http://dx.doi.org/10.1016/j.athoracsur.2006.10.069.

41. Hövels-Gürich HH, Konrad K, Skorzenski D. Long-term behavior and quality of life after corrective cardiac surgery in infancy for tetralogy of Fallot or ventricular septal defect. *Pediatr Cardiol.* 2007;28:346–354.

42. Hövels-Gürich HH, Konrad K, Skorzenski D, et al. Long-term neurodevelopmental outcome and exercise capacity after corrective surgery for tetralogy of Fallot or ventricular septal defect in infancy. *Ann Thorac Surg.* 2006;81(3):958–966. http://dx.doi.org/10.1016/j.athoracsur.2005.09.010.

43. Hövels-Gürich HH, Bauer SB, Schnitker R, et al. Long-term outcome of speech and language in children after corrective surgery for cyanotic or acyanotic cardiac defects in infancy. *Eur J Paediatr Neurol*. 2008;12(5):378–386.

44. Shillingford AJ, Glanzman MM, Ittenbach RF. Inattention, hyperactivity, and school performance in a population of school-age children with complex congenital heart disease. *Pediatrics*. 2008;121(4):e759–767.

45. Achenbach TM, Rescorla L. *Manual for the ASEBA School-age Forms & Profiles*; 2001.

46. Rueda MR, Fan J, McCandliss BD, Halparin JD. Development of attentional networks in childhood. *Neuropsychologia*. 2004;42(8):1029–1040.

47. Schilling F, Kiphard EJ. *Körperkoordinationstest Für Kinder*; 1974.

48. Beery KE. *The VMI, Developmental Test of Visual-Motor Integration*; 1989.

49. Mahle WT, Clancy RR, Moss EM, Gerdes M, Jobes DR, Wernovsky G. Neurodevelopmental outcome and lifestyle assessment in school-aged. *Pediatrics*. 2000;105(5):1082–1089.

50. Achenbach TM. *Manual for the Child Behaviour Checklist/4–18 and 1991 Profile*. Burlington, VT: University of Vermont Department of Psychiatry; 1991.

51. Snookes SH, Gunn JK, Eldridge BJ, Donath SM. A systematic review of motor and cognitive outcomes after early surgery for congenital heart disease. *Pediatrics*. 2010;125(4):e818–e827.

52. Fair DA, Bathula D, Nikolas MA, Nigg JT. Distinct neuropsychological subgroups in typically developing youth inform heterogeneity in children with ADHD. *Proc Natl Acad Sci*. 2012;109(17):6769–6774.

53. Lavie N. Distracted and confused?: selective attention under load. *Trends Cogn Sci*. 2005;9(2):75–82. http://dx.doi.org/10.1016/j.tics.2004.12.004.

54. McNab F, Klingberg T. Prefrontal cortex and basal ganglia control access to working memory. *Nat Neurosci*. 2007;11(1):103–107. http://dx.doi.org/10.1038/nn2024.

55. Beaumont JL, Havlik R, Cook KF, et al. Norming plans for the NIH toolbox. *Neurology*. 2013;80(11):S87–S92.

56. Dunbar-Masterson C, Wypij D, Bellinger DC. General health status of children with D-transposition of the great arteries after the arterial switch operation. *Circulation*. 2001;104:I138–I142.

57. Williams DL, Gelijns AC, Moskowitz AJ. Hypoplastic left heart syndrome: valuing the survival. *J Thorac Cardiovasc Surg*. 2000;119(4 Pt 1):720–731.

58. Drotar D, Stancin T, Dworkin PH, Sices L, Wood S. Selecting developmental surveillance and screening tools. *Pediatr Rev*. 2008;29(10):e52–e58. http://dx.doi.org/10.1542/pir.29-10-e52.

59. Marino BS, Beebe D, Cassedy A, Riedel M. Executive functioning, gross motor ability and mood are key drivers of poorer quality of life in child and adolescent survivors with complex congenital heart disease. *J Am Coll Cardiol*. 2011;57(14):E421.

60. Miyake A, Friedman NP, Emerson MJ, Witzki AH. The unity and diversity of executive functions and their contributions to complex "frontal lobe" tasks: a latent variable analysis. *Cogn Psychol*. 2000;41:49–100.

61. Cepeda NJ, Blackwell KA, Munakata Y. Speed isn't everything: complex processing speed measures mask individual differences and developmental changes in executive control. *Dev Sci*. 2013;16(2):269–286.

62. Rovee CK, Rovee DT. Conjugate reinforcement of infant exploratory behavior. *J Exp Child Psychol*. 1969;8(1):33–39.

63. Ireton H, Thwing E. *Minnesota Child Development Inventory*; 1974.

64. David N, Bricker DD, Paquin J, et al. *Ages and Stages Questionnaires (ASQ)*. Brookes Publishing Company; 1999.

65. Wechsler D. *Wechsler Preschool and Primary Scale of Intelligence—Revised UK Edition*. New York: The Psychological Corporation; 1989.

66. Korkman M, Kemp S, Kirk U. *Nepsy*; 1998.

67. Wechsler D. *Wechsler Preschool and Primary Scale of Intelligence.* 3rd ed. San Antonio, TX: Harcourt Assessment Company; 2002.
68. Woodcock RW, McGrew KS, Mather N. *Woodcock-Johnson Achievement Battery III, Manual;* 2001.
69. McGoey KE, DuPaul GJ, Haley E, Shelton TL. Parent and teacher ratings of attention-deficit/hyperactivity disorder in preschool: the ADHD rating scale-IV preschool version. *J Psychopathol Behav Assess.* 2007;29(4):269–276.
70. Merrell K. *Preschool and Kindergarten Behavior Scales.* Austin, TX: Clinical Psychology Publishing Company, Inc.; 1994.
71. Achenbach TM, Rescorla LA. *Child Behavior Checklist for Ages 1(1/2)–5;* 2000.
72. Folio MR, Fewell RR. *Peabody Developmental Motor Scales and Activity Cards;* 1983.
73. Reitan RM, Davison LA. *Clinical Neuropsychology: Current Status and Applications.* Halsted Press; 1974.
74. Carrow-Woolfolk E. *Test for Auditory Comprehension of Language.* DLM Teaching Resources; 1985.
75. Gardner MF. *Receptive One-Word Picture Vocabulary Test 2000;* 1985.
76. Gardner MF. *Expressive One-Word Picture Vocabulary Test;* 1990.
77. Kirk S, McCarthy J, Kirk W. *Illinois Test of Psycholinguistic Abilities (Revised).* Urbana, IL: University of Illinois Press; 1968.
78. Scarborough HS. Index of productive syntax. *Appl Psycholinguistics.* 1990;11(01):1–22. http://dx.doi.org/10.1017/S0142716400008262.
79. Robbins J, Klee T. Clinical assessment of oropharyngeal motor development in young children. *J Speech Hear Disord.* 1987;52(3):271–277. http://dx.doi.org/10.1044/jshd.5203.271.
80. Darley FL, Aronson AE, Brown JR. *Motor Speech Disorders.* Boston, MA: Little Brown & Co.; 1975.
81. Goldman R, Fristoe M. *Goldman-Fristoe Test of Articulation.* American Guidance Service, Inc.; 1972.
82. Wechsler D. *Wechsler Intelligence Scale for Children.* 3rd ed. San Antonio, TX: The Psychological Corporation; 1991.
83. Klove H. Clinical neuropsychology. *Med Clin North Am.* 1963;47:1647–1658.
84. Semel E, Wiig E, Secord W. *Clinical Evaluation of Language Fundamentals;* 1995.
85. Spreen O, Strauss E. *A Compendium of Neuropsychological Tests: Administration, Norms, Commentary.* Oxford University Press; 2008.
86. MacCarthy D. *Manual for the McCarthy Scales of Children's Abilities;* 1972.
87. Rosner J. *Phonic Analysis Training and Beginning Reading Skills;* 1971.
88. Sheslow D, Adams W. *Wide Range Assessment of Memory and Learning (WRAML);* 1990.
89. Heaton RK, Chelune GJ, Talley JL, Kay GG, Curtiss G. *Wisconsin Card Sorting Test. Revised and Expanded.* Psychological Assessment Resources; 1993.
90. Marcovitch S. *The Development of Executive Function in Early Childhood;* 2003.
91. Beery KE. *The Beery-Buktenica Developmental Test of Visual-Motor Integration: Admin-Istration, Scoring and Teaching Manual.* 4th ed. Modern Curriculum Press; 1997.
92. Bernstein JH, Waber DP. *Developmental Scoring System for the Rey-Osterrieth Complex Figure.* Odessa, FL: Psychological Assessment Resources, Inc.; 1996.
93. Wechsler D. *Wechsler Individual Achievement Test Manual.* San Antonio, TX: Psychological Cor; 1992.
94. Achenbach TM. *Manual for the Teacher's Report Form and 1991 Profile.* Univ Vermont/Department Psychiatry; 1991.
95. Wechsler D. *Wechsler Abbreviated Scale of Intelligence WASI;* 1999.
96. Buschke H. Selective reminding for analysis of memory and learning. *J Verbal Learn Verbal Behav.* 1973;12(5):543–550.
97. Benton AL, Sivan AB. *Benton Visual Retention Test;* 1992.
98. Cohen MJ. *Children's Memory Scale.* San Antonio, TX: The Psychological Corporation; 1997.

99. Culbertson WC, Zillmer EA. *Tower of London- Drexel University (TOLDX) Technical Manual*. North Tonawanda, NY: Multi-Health Systems; 2001.
100. Reynolds CR, Kamphaus RW. *BASC*. Circle Pines, MN: American Guidance Service; 1992.
101. DuPaul GJ, Powers TJ, Anastopoulos AD, Reid R. *The ADHD Rating Scale IV Manual*. New York, NY: Guilford Press; 1999.
102. Melchers P, Kaufman AS, Kaufman NL. *Kaufman-Assessment Battery for Children, Deutsche Version (K-ABC)*; 1994.
103. Rosner J. Screening for perceptual skills dysfunction: an up-date. *J Am Optom Assoc.* 1979;50(10):1115–1119.
104. Corp P. *The Wechsler Individual Achievement Test*. 2nd ed. San Antonio, TX: The Psychological Corporation; 2002.
105. Conners CK. *Conners' Rating Scales Revised*. Multi-Health Systems; 2001.
106. Delis DC, Kaplan E, Kramer JH. *Delis-Kaplan Executive Function System*; 2001.
107. Gioia GA. *Behavior Rating Inventory of Executive Function*; 2000.
108. Guy SC, Isquith PK, Gioia GA. *BRIEF-SR*; 2004.
109. Gardner MF. *Test of Visual-Perceptual Skills (Non-Motor), Revised*. Hydesville, CA: Psychological and Educational Publications, Inc.; 1997.
110. Holmes JH, Waber DP. *Developmental Scoring System for the Rey-Osterrieth Complex Figure*. Odessa, FL: Psychological Assessment Resources, Inc.; 1996.
111. Hegarty M, Richardson AE, Montello DR, Lovelace K. Development of a self-report measure of environmental spatial ability. *Intelligence.* 2002;30(5):425–447.
112. Cohen SB, Wheelwright S, Hill J. The "Reading the Mind in the Eyes" test revised version: a study with normal adults, and adults with Asperger syndrome or high-functioning autism. *J Child Psychol Psychiatry.* 2001;42(2):241–251.

PART III

PSYCHOLOGICAL PROFILES AND PROCESSES

Is There a Behavioral Phenotype for Children With Congenital Heart Disease?

C. McCusker

The Queens University of Belfast, Belfast, Northern Ireland; The Royal Belfast Hospital for Sick Children, Belfast, Northern Ireland

F. Casey

The Royal Belfast Hospital for Sick Children, Belfast, Northern Ireland

OUTLINE

In Section 2 we looked at evidence that a neurodevelopmental phenotype is emerging in the literature for children with significant congenital heart disease (CHD). As with all phenotypes, individual variation will exist due to the range of interacting factors involved. However, patterns are suggested in terms of early motor deficits and later difficulties in the domains of attention and executive functioning. In this chapter we look at whether similar patterns are becoming evident in terms of emotional and behavioral outcomes. This is an important question to pursue. Not only do such outcomes interfere with adaptive functioning through childhood, building cumulative risk, but we also know there is significant continuity between childhood and adult adjustment in general and for children and adults with CHD in particular.[1] Early intervention is recognized as both clinically and socially important for children and families at risk in general.[2] Such efforts are likely to be most effective and efficient if tailored to the specific needs and presentation of the targeted group.

POTENTIAL PATHWAYS TO BEHAVIORAL MALADJUSTMENT

As noted in the chapter "Historical Perspectives in Pediatric Psychology and Congenital Heart Disease," the emergence of pediatric psychology as a discipline preceded much of any extensive evidence base, and was at least partly promoted by clinical recognition of how chronic illness and disability could pose challenges for normal childhood development.[3,4] In this section we consider how CHD can challenge psychological pathways to healthy childhood adjustment.

Bowlby and other early theorists in infant mental health[5,6] highlighted the importance of a secure mother–infant attachment for the subsequent cognitive, emotional, and behavioral development of the child. Indeed poor attachments, putatively formed in infancy and early childhood, have increasingly been implicated in a wide range of child and adult psychopathologies.[7] In CHD a catalog of factors challenge healthy attachments. Breathlessness, difficulties feeding, and difficult temperaments[8,9] coupled with the neurodevelopmental features outlined in the last section, compromise the infants' capacity to play their part in the reciprocal behaviors which promote attachment. The fragile infant is typically separated from the mother, in a stressful medical environment, for prolonged periods of time and the very caregiving interactions, crucial to attachment formation, can be significantly reduced. Maternal responsivity, as well as capacity to "buffer" these adverse influences, is additionally compromised by the stress and elevated levels of emotional difficulties associated with the diagnosis, invasive interventions, and uncertain future.[10]

However, potential pathways to maladjustment are not restricted to infancy. Physical limitations, both due to the neurodevelopmental deficits—noted in the preceding chapters—and due to reduced stamina may compromise the child's capacity to engage in play and physical activities, which promote social skills and psychological adjustment.[11] Parental perceptions are likely to be just as powerful here as actual limitations.

Repeated invasive procedures and hospitalizations, together with associated separation from the family, have the potential to be traumatic and distressing for the child.[12] Given such a context it is unsurprising that parents may "spoil" the child, with normal discipline processes being compromised by guilt, worry, and fear that conflict may actually damage the heart. Inconsistency in parenting appears likely as parents may oscillate between overprotection and discipline, which is a recipe for behavioral difficulties in the child.

Such potential pathways to psychological maladjustment continue across childhood and into adolescence. Typically, with age the children's own perceptions of their presumed fragility, of having a heart condition, of differences with their peers, of their experience of exclusion from, or limitations in, physical activities start to become more preponderant and can set up additional processes which can potentiate emotional, peer relationship, and other behavioral difficulties. Knowledge and understanding is typically limited[13] at a time when thoughts may be turning to establishing independence, romantic relationships, and future career and parenting potential.

Considering all these influences, which one might expect to act cumulatively as suggested by models of developmental psychopathology,[14] one might also expect to see very high prevalence rates of emotional and behavioral disturbances within these children and young people. However, the first key message of this chapter is that this is not the case.

NOISE IN THE EVIDENCE BASE

Progressing knowledge by research is rarely linear in its trajectory. Methodological variation in sampling, assessment tools, analytical procedures, and general approaches to internal and external validity will obscure comparability of findings.[15] In addition, a cohort of children studied today may be very different from cohorts studied during the 1980s and 1990s due to advances in surgical procedures, hospital, and general psychosocial care. Thus even hypothetically "perfect" methodologies may yield different conclusions. To compound matters further, publication bias[16] may reflect a distorted picture of outcomes. Such a context creates "noise" in the emergent evidence base and makes it challenging to synthesize what is important and evidential from what is artifactual or misleading.

If we consider the background to our own work to be reported in this chapter, contradictions and problems with the evidence base for children with CHD are apparent. Thus, while many studies have suggested an increased prevalence of emotional and behavioral disturbances in children with CHD, others have found no differences in comparison to controls or norms.[17–19] Indeed, on some indices of psychological adjustment, outcomes have actually been reported as better than with controls.[20] Such mixed findings are not uncommon in the general pediatric literature,[21] and it is important to recognize that a *successful* negotiation of the potential pathways to maladjustment, outlined above, may in fact confer resilience rather than risk.[22] Evaluating outcomes for children with CHD is not a simple process and models are likely to be multifactorial.[23] If these are not taken into account contradictory and misleading findings will be inevitable.

First, despite our knowing since De Maso's research during the early 1990s that maternal perceptions in relation to their child's heart disease may be more important than disease severity per se,[24] maternal and family factors have been neglected until recently in research models. It is likely that outcomes for these children will not vary homogeneously, but rather interact with family, social, personal, disease, and treatment factors. Moreover, interactions may very well impact in different ways on different domains of functioning (eg, emotional difficulties vs social competencies and home vs school presentations).

Secondly, but related, too often the research has relied on single informants (usually mothers), even though we have known that maternal reports can be at odds with those reported by young people themselves and independent teacher ratings.[17,25,26] Indeed at school it can sometimes appear that such problems are not at all in evidence[27] which must raise questions with respect to situational specificity at least, and reliability of maternal reports at worst—especially given the fact that elevated mood difficulties in mothers may bias perceptual judgments.

Thirdly, while we use and endorse the value of the *Child Behavior Checklist* (CBCL)[28] as an internationally recognized, psychometrically sound, differential, and multidimensional measure of childhood adjustment across different domains of functioning, we have also highlighted how an indiscriminating use of this tool has produced misleading results. Specifically, we have shown that when items normally related to mood disturbance, withdrawn, somatic, and internalizing behaviors are removed, as they in fact also relate to physical features of the disease (eg, "underactive … dizzy … tired … headaches"), then elevated scores and profiles on these "internalizing" dimensions are reduced or actually disappear.[29] Unfortunately, this has not always been accounted for, and thus reported rates of elevated "internalizing" behaviors may be misleading—not just with the CHD population but in the pediatric population more generally.

Finally, we have drawn attention to other possible methodological sources of bias in the literature.[30,31] Work has often been cross-sectional in nature which makes inferences about cause and effect between different factors and outcomes difficult—especially when some of those factors assumed to be "predictive" (eg, family factors) might also themselves be influenced by the putative outcome measure (ie, child adjustment). Prospective studies are rare—with the *Boston Circulatory Arrest Study* being seminal in this regard (see chapter: A Longitudinal Study From Infancy to Adolescence of the Neurodevelopmental Phenotype Associated With d-Transposition of the Great Arteries). However, these often include studies with a single cardiac disease (eg, transposition of the great arteries) or with a class of cardiac diseases (eg, cyanotic defects), which can be a problem if behavioral outcomes actually vary across conditions.

Given these challenges to interpretation of findings, how might we move toward synthesis and conclusion? First, we can look at the conclusions of systematic reviews and meta-analyses, which synthesize findings from across a number of studies which meet some key standards of quality and present their data in such a way that comparisons across studies can be made. The results of such studies involving children with CHD is reasonably comparable to those with chronic illnesses in general in suggesting small to medium effect sizes for elevated rates of behavior problems in the clinical samples in comparison to controls.[17,21,32] Put another way, these children appear over twice as likely to experience emotional and behavioral difficulties as their healthy peers. However, these summative reviews are only as good as the methodologies of the historical studies included and effect sizes (derived from parental report data) are markedly reduced in those studies involving other informants. Moreover, amalgamating data in this way risks obscuring important methodological differences and associated effect size variations which might actually illuminate conclusions. Arguably, the systematic and meta-analytic methodology is premature when models of understanding are in the process of being built—as opposed to tested—as in this case.[33] A more critical and interpretive review is called for. This is put forward in the remainder of this chapter by doing two things. First, we describe a contemporary program of longitudinal research at our center which has been informed by many of the methodological issues outlined above. Secondly, we engage in a critical interpretive synthesis[33] by looking for patterns across other contemporary studies, which have been similarly informed, and offering interpretations of phenotypical meaning therein.

THE CONGENITAL HEART DISEASE INTERVENTION PROGRAM

The Congenital Heart disease Intervention Program (CHIP) began in Belfast, Northern Ireland, during the early 2000s. This represented a multiprofessional collaboration among clinical psychologists, pediatric cardiologists, nurses, pediatricians, and a service user organization. Its ultimate purpose was to formulate and implement a secondary prevention program of psychological interventions to improve adjustment in children and families with CHD. It has been proposed that targeting interventions at key developmental transitions is likely to be most effective,[34] and our programs (described in chapters: The Congenital Heart Disease Intervention Program (CHIP) and Interventions in Infancy; Growing up – Interventions in Childhood and CHIP-School) first recruited families in infancy (at diagnosis and initial surgeries) and in early childhood—as the child was starting school. However, in addition to these intervention programs and their evaluation, we have been able to establish one of the most comprehensive and longitudinal data sets of outcomes for children with a spectrum of significant CHD presentations. Our longitudinal work here is ongoing; however, our current findings in terms of neurodevelopmental outcomes are discussed in chapter "Neurodevelopmental Patterns in Congenital Heart Disease Across Childhood: Longitudinal Studies From Europe." In this chapter the current picture in terms of psychosocial outcomes are described together with the factors we found to be important in predicting outcome.

Behavioral Outcomes: Studies 1 and 2

In these first studies we assessed 90 children (4–5 years of age) who had undergone at least one invasive procedure for correction or palliation of a major heart defect in infancy.[30,35] Children with diagnosed developmental or psychiatric syndromes were excluded, but the sample included those with a comprehensive range of acyanotic and cyanotic heart conditions (see chapter: Congenital Heart Disease: The Evolution of Diagnosis, Treatments, and Outcomes) and who had undergone cardiopulmonary bypass, or "open" repairs and palliation as well as transcatheter or "closed" interventions. We incorporated a specific control group here, rather than solely relying on reference group norms. This was a group of children who had been diagnosed with mild, self-correcting, heart defects in infancy, matched for group demographics.

Rates of behavior problems (as indexed by the CBCL) were not universally higher in the CHD groups and indeed only 5% of those with corrected cyanotic conditions were in the clinically significant range compared with 16% of the control group. However, for acyanotic conditions (whether corrected by open or closed interventions) rates were higher (20–21%) and for the children with complex/palliated conditions rates were significantly higher (33%).

Of greater interest at this time was an examination of factors which determined outcomes. While disease and surgical factors such as length of time on bypass had predicted neurodevelopmental outcomes for this same sample (see chapter Neurodevelopmental Patterns in Congenital Heart Disease Across Childhood: Longitudinal Studies From Europe) their contribution to behavioral outcomes (as reported by the parents) was low. Rather, a hugely significant 59% of the variance on behavioral outcomes was explained by maternal and family factors (maternal worry, maternal mental health, parenting style, and family composition). Even more striking were associations with levels of physical activity and unscheduled medical consultations taken to reflect health anxiety. Activity levels, which have been related to behavioral adjustment in children, were just as influenced by maternal worry as severity of the condition. Surprisingly, unscheduled medical consultations were actually lowest of all in the most severe, complex CHD, subgroup. On this variable, maternal worry and social deprivation were the primary predictors of outcome.

Although we had an apposite control group, and a range of cardiac presentations, these studies were essentially cross-sectional in nature and relied on maternal reports only. A third, longitudinal, study would take forward and refine conclusions.

Behavioral Outcomes: Study 3

In this study we followed up the 70 infants and families from the CHIP project who had been assessed in the first year of life, 7 years later.[31] Children had a similar range of cardiac presentations as described in our previous studies. However, this longitudinal design was more robust in determining associations between predictor factors assessed in infancy and outcomes 7 years later. In addition, we looked at behavioral outcomes as reported by parents, as well as by independent teacher ratings. Finally, we incorporated a stringent control group not previously utilized in this literature—nearest age healthy siblings. Such a group controls for demographic, as well as nonspecific family, factors.

Excluding children with development syndromes (eg, Down syndrome) from our analyses, given their own phenotypical presentations, and amalgamating specific CHD subgroups (as no subgroup differences were detected on preliminary analyses in this sample), the outcomes on the CBCL are summarized in Table 6.1.

Findings were both statistically and clinically significant. Three main conclusions are suggested. First, in comparison to a sibling control group, rates of behavior problems were elevated and competencies reduced, as evaluated by *both* parents and independent teacher raters. Secondly, although still significantly different, the magnitude of the differences with siblings was distinctly less when rated by teachers than parents (27% vs 7% in the clinically significant range of the behavior problem scales and 32% vs 15% on the competence scales). Thirdly, despite the absolute differences between parental and teacher ratings the pattern was the same (ie, more deficits in school, social, and activity competencies than with behavior problems per se).

TABLE 6.1   Mean (Standard Deviation) Child Behavior Checklist Problems and Competence Scores as Rated by Parents and Teachers Independently

	CHD[a] ($n=31$)	Siblings ($n=18$)	F (df)	P	[b]Partial η^2
Parents					
Total problems (SD)	51.2 (11.3)	42.9 (8.9)	6.8 (1.46)	0.012	0.13
95% CIs	47.2–55.1	37.9–47.9			
% Clin. Sig.[c]	27%	0%			
Total competence (SD)	41.2 (10.9)	51.0 (10.8)	8.6 (1.43)	0.005	0.17
95% CIs	37.1–45.3	45.7–56.3			
% Clin. Sig.[c]	32%	18%			
Teachers					
Total problems (SD)	47.0 (10.0)	40.5 (6.8)	4.2 (1.37)	0.048	0.10
95% CIs	43.4–50.6	35.1–45.8			
% Clin. Sig.[c]	7%	0%			
Adaptive functioning (SD)	48.5 (7.4)	54.7 (5.4)	6.7 (1.37)	0.014	0.15
95% CIs	45.9–51.2	50.7–58.7			
% Clin. Sig.[c]	15%	0%			

[a] *Congenital Heart Disease group.*
[b] *Partial eta squared effect size: >0.08 = medium effect size; >0.13 = large effect size.*
[c] *Percentage of cases in the clinically significant range.*

Furthermore, when we unpacked the total index scores summarized in Table 6.1 and excluded the somatic scale items, for the reasons outlined above, the subscales which particularly defined the CHD children's difficulties were in relation to *social problems, thought problems, and attention problems.*[31] Although levels of statistical significance varied, it was striking that the clinical effect sizes were greatest in these domains as rated by both parents and teachers, thus enhancing our confidence in the credibility and validity of this pattern. Thus, even within the behavior problems scale of the CBCL, problems appear related to personal and interpersonal behaviors or competencies—rather than the sorts of "internalizing" problems with anxiety, depression, and withdrawal which have been traditionally associated with this population. Related, our teacher reports also suggested lower levels of participation in physical education activities and a greater number of days missed from school in the CHD children in comparison with their siblings.[31]

Turning toward mediating mechanisms, Table 6.2 summarizes the final regression models for outcomes on the problems and competences scales

TABLE 6.2   Final Regression Models of Disease/Surgical, Child, and Family Factors and Standardized Betas Associated With Child Behavior Checklist Outcomes

	Parents		Teachers	
	Total Problems	Total Competence	Total Problems	Adaptive Functioning
Disease/Surgical				
Cyanosis		0.468		
Open surgery			0.177	
Length of time on bypass		−0.524		
Syndrome			−0.273	
Child				
Gender		−0.775		−0.279
Bayley's MDI[a] (8 months)		0.170	−0.373	
Bayley's PDI[b] (8 months)			−0.288	0.471
Family				
Lone parent	−0.282			
Maternal GSI[c] score (8 months)	0.357			−0.355
Family cohesion[d]	−0.162		−0.435	
Adjusted R^2	0.331	0.396	0.415	0.551
F (p)	5.79 (0.004)	3.46 (0.046)	4.83 (0.004)	12.46 (<0.001)

[a] Mental Development Index—Bayley Scales of Infant Development.[36]
[b] Psychomotor Development Index—Bayley Scales of Infant Development.[36]
[c] General Severity Index score on the Brief Symptom Inventory[37] of psychopathology.
[d] Family Cohesion—scale from the Family Environment Scale.[38]

of the CBCL as rated by both parents and teachers. Two main conclusions are possible:

- These prospective analyses confirmed the significance of maternal and family factors over disease and surgical factors in determining behavioral problems for these children. *Only* these factors, and not disease or surgical factors, were implicated in long-term behavioral outcomes as based on parental reports. Even on the teacher scales, where neurodevelopmental, disease, and surgical factors became important, a family factor—*family cohesion* exerted the greatest influence.
- However, outcomes on the competence scales (related to activities, social and school functioning) were not associated with these maternal

and family factors. Rather, here we found what might be the more expected involvement of disease factors (having a cyanotic condition), surgical factors (length of time on bypass), and child factors (having more neurodevelopmental deficits in infancy and being male) on these long-term outcomes.

- At school, and as based on teacher reports, a combination of factors appeared relevant to understanding both behavior difficulties and adaptive functioning (competences), although maternal and family factors and neurodevelopmental status in infancy were of greatest significance.

TOWARD A SYNTHESIS OF PHENOTYPICAL PRESENTATION AND PROCESSES

Elevated risk for adjustment difficulties (behavior problems and competences) have been confirmed in our work with children with significant CHD in comparison to stringent control groups—children with mild defects[30] and with siblings.[31] While family factors have featured significantly in our studies as mediating outcome, the fact that we found significant levels of adjustment difficulties in children with CHD in comparison to their siblings highlights that it is indeed the disease status which sets the stage for these later difficulties. In common with other studies during 2009–10,[27,39] lower absolute levels of behavior problems were reported by teachers at school. However, the finding that differences, in comparison to siblings, remain significant on teacher ratings at school, suggests that this is not just about biased maternal perceptions or indeed is limited just to the family environment.

Despite variations in absolute levels, the pattern of adjustment difficulties was consistent across both parent and independent teacher reports. When we control for the confounding effect of somatic items on behavioral scales, these children (at least in middle childhood) do not appear to be troubled by elevated levels of mood disturbances per se. Rather, the behavioral phenotype appears to be particularly related to personal and interpersonal competencies. Thus school competencies (related to academic achievement, activity levels, and social functioning) were reported by both parents and teachers. Moreover, those behavior problems which were most elevated—again on both parent and teacher reports—related to the CBCL subscales of *attention* (inattention and hyperactivity), *social problems* (peer relationship skills), and *thought problems* (ritualistic behaviors, sleep disturbance, and idiosyncratic thought patterns).

Consistent with this conclusion, such patterns have been increasingly reported in work from other centers. The *Boston Circulatory Arrest Study*[39] also found a greater preponderance of problems with personal and interpersonal competencies, including thought, attention, and social problems, rather than emotional problems per se. Indeed Bellinger has suggested that deficits in social cognition or theory of mind, possibly at least partially related to the executive functioning difficulties (as described in chapters: A Longitudinal Study From Infancy to Adolescence of the Neurodevelopmental Phenotype Associated With d-Transposition of the Great Arteries, Neurodevelopmental Patterns in Congenital Heart Disease Across Childhood: Longitudinal Studies From Europe, and An Emergent Phenotype: A Critical Review of Neurodevelopmental Outcomes for Complex Congenital Heart Disease Survivors During Infancy, Childhood, and Adolescence), may underlie these interpersonal competencies.[40] Other European centers have similarly found (1997, 2004, and 2007) elevated levels of attention, social, and thought problems,[41–43] with increased levels of autistic spectrum, and attention deficit, disorders being reported in 2014 in this population.[44]

It is clear from studies with teenagers and indeed adults that these problems do not go away, and if anything, increase with age, certainly across childhood.[17,45] Moreover, a longitudinal Dutch study in 2005 found that those in the clinically significant range for emotional and behavioral problems on the CBCL in adolescence were 3.4 times more likely to be in the clinically significant range of parallel scales in adulthood.[46]

As noted at the outset of this chapter, early intervention promises to be important in reducing the personal, social, and indeed, economic costs of these problems. This needs to be tailored to the specific nature of difficulties encountered, as well as by an understanding of the factors which reduce or exacerbate risk to target interventions. The results from our studies, in combination with others in the area, point toward a multifactorial model which includes disease, surgical, child, and family factors as summarized in Fig. 6.1.

Increasingly, across studies, outcomes have not been shown to be primarily mediated by the severity of the heart defect itself.[17,30,31,47] Moreover, despite advances in surgical techniques and medical care, analyses have suggested comparable levels of behavioral disturbance in cohorts from the 1980s and mid-1990s.[48] The heart disease may very well set the stage for difficulties with these children, but it seems that many other factors determine the evolution and extent of these.

Our studies have found that maternal and family factors play a highly significant role in determining behavioral outcomes. This not only included levels of behavioral adjustment, which might be understandable, but also outcomes which might be thought to be more related to the disease severity, such as activity levels and health concerns (as indexed by frequency of unscheduled medical consultations). However, if we review the potential pathways to maladjustment outlined at the start of this chapter, it is clear that maternal and family resources can serve a protective function, or an undermining force if such are poor. Thus poor mental health, worry,

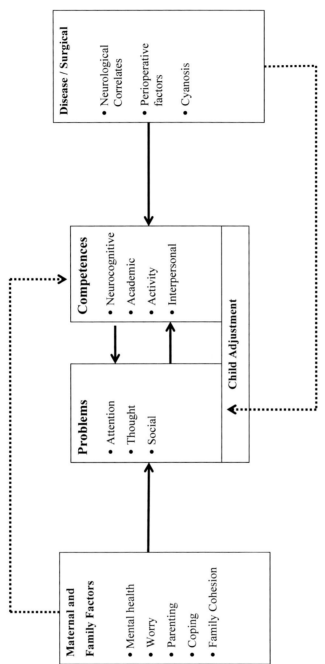

FIGURE 6.1 Key factors in understanding the behavioral phenotype of children with congenital heart disease.

limited parenting skills, and a family context lacking in cohesion are all likely to exacerbate the challenges the heart defect brings to everything from forming a good attachment with the infant to promoting optimal independence, academic and social inclusion, and behavioral adjustment in the child. Unfortunately, as discussed by Doherty and Utens in the chapter "A Family Affair," maternal and family functioning is challenged in these families. Fortunately, as discussed in the next section of this book, these predictive associations do point toward new possibilities for intervention—which take a family and systemic perspective.

While maternal and family factors are seen as central in the model outlined in Fig. 6.1, an interacting multifactorial perspective is required. Thus, behavioral adjustment is likely to be influenced by personal competences in the child. As outlined elsewhere in this volume, neurodevelopmental deficits are common in these children and have been shown to be related to behavioral outcomes.[25,31,39] Such cognitive deficits may underlie the social problem-solving skills fundamental to negotiating personal and interpersonal development as discussed above. In turn, surgical factors such as length of time on cardiopulmonary bypass have been shown to be important predictors of neurodevelopmental outcomes which may in turn explain why this has also sometimes been associated with behavioral outcomes[25,31] (ie, via an indirect pathway as well as via direct impact on neurological systems which are important in behavior regulation).

Undoubtedly, other factors such as comorbidity, gender, socioeconomic status, and factors related to school and medical care systems are likely to be relevant. These have not yet been so well researched or reliably reported in the literature, but they may very well occur. Thus the model depicted in Fig. 6.1 does not purport to be definitive or fully comprehensive. However, it represents the best fit model which accommodates research data at the present time. In sum, we propose that congenital heart disease sets the context for childhood difficulties, through direct and indirect pathways as outlined above, and where maternal and family factors, neurological correlates, and the impact of bypass surgery on neurodevelopmental resources are of primary influence.

LOOKING TOWARD INTERVENTION

Given what we know about risk for neurodevelopmental deficits and the likely stress associated with repeated invasive cardiac interventions, clinical consensus statements have appeared in the literature with respect to neurodevelopmental surveillance and preparation of the child and family for such cardiac procedures.[49,50] These are undoubtedly warranted and are impressive in the detailed consideration given to assessment, remedial, and therapeutic interventions possible.

However, until the CHIP program, there had been no exposition and evaluation of a systematic and comprehensive intervention program for these children and their families. The etiological research outlined in this chapter, and indeed the previous section on neurodevelopmental outcomes, point toward the relevance of a family-focused, systemic, intervention. While medical and surgical advances will have their part to play in the future, we have noted earlier that such advances to date have not appreciably changed psychosocial outcomes for these children across time. Psychosocial research suggests another direction of travel here. Specifically it points toward the value of competence-promoting interventions for these children (rather than targeting mood disturbance per se), and it highlights the central importance of working collaboratively with mothers and families to bolster their own personal resources to affect this. This is the focus of the final section of this book.

References

1. Van Rijen EH, Utens EM, Roos-Hesselink JW, et al. Psychosocial functioning of the adult with congenital heart disease: a 20–33 years follow-up. *Eur Heart J.* 2003;24:673–683.
2. Shonkoff J, Phillips D. *From Neurons to Neighbourhoods: The Science of Early Childhood Development.* Washington, DC: National Academy Press; 2000.
3. Aylward BS, Bender JA, Graves MM, Roberts M. Historical developments and trends in pediatric psychology. In: Roberts M, Steele RC, eds. *Handbook of Pediatric Psychology.* 4th ed. New York: Guildford Press; 2009.
4. Eiser C. *Chronic Childhood Disease: An Introduction to Psychological Theory and Research.* Cambridge University press; 1990.
5. Bowlby J. The nature of the child's tie to his mother. *Int J Psychoanal.* 1958;39:350–373.
6. Sameroff AJ. Developmental systems and psychopathology. *Dev Psychopathol.* 2000;12:297–312.
7. Sroufe A. Attachment and development: a prospective, longitudinal study from birth to adulthood. *Attach Hum Dev.* 2005;7:349–367.
8. Clemente C, Barnes J, Shinebourne E, Stein A. Are infant behavioural difficulties associated with congenital heart disease? *Child Care Health Dev.* 2001;27:47–59.
9. Sterne-Larsen K, Brandlistuen R, Holmstrom H, Landolt M, Eskedal L, Vollrath M. Emotional reactivity in infants with congenital heart defects: findings from a large case-cohort study in Norway. *Acta Paediatr.* 2010;99:52–55.
10. Cousino MK, Hazen R. Parenting stress among caregivers of children with chronic illness: a systematic review. *J Pediatr Psychol.* 2013;38:809–828.
11. Wallander JL, Varni JW. Effects of pediatric chronic physical disorders on child and family adjustment. *J Child Psychol Psychiatry.* 1998;39:29–46.
12. Garson SL. Psychological aspects of heart disease in childhood. In: Garson A, Bricker JT, Fishes DT, Neishe SR, eds. *The Science and Practice of Pediatric Cardiology.* 2nd ed. Baltimore: Williams & Wilkins; 1998.
13. Tong EM, Kools S. Health care transitions for adolescents with congenital heart disease: patient and family perspectives. *Nurs Clin North Am.* 2004;18(2):93–98.
14. Rutter M, Sroufe A. Developmental psychopathology: concepts and challenges. *Dev Psychopath.* 2000;3:265–296.
15. Kuhn TS. *The Structure of Scientific Revolutions.* Chicago: The University of Chicago Press; 2012.
16. Gluud LL. Bias in clinical intervention research. *Am J Epidemiol.* 2006;163:493–501.

17. Karsdorp P, Everaerd W, Kindt M, Mulder B. Psychological and cognitive functioning in children and adolescents with congenital heart disease: a meta-analysis. *J Pediatr Psychol.* 2007;32(5):527–541.

18. Menahmen S, Poulakis Z, Prior M. Children subjected to cardiac surgery for congenital heart disease. Part 1 – emotional and psychological outcomes. *Interact Cardiovasc Thorac Surg.* 2008;7:600–604.

19. Utens E, Verslusis-Den Bieman H, Witsenburg M, Bogers AJ, Verhulst FC, Hess J. Cognitive and behavioural and emotional functioning of young children awaiting elective cardiac surgery or catheter intervention. *Cardiol Young.* 2001;11:153–160.

20. Salzer-Mohar U, Herle M, Floquet P, et al. Self-concept in male and female adolescents with congenital heart disease. *Clin Pediatr.* 2002;41:17–24.

21. Barlow JH, Ellard DR. The psychosocial well-being of children with chronic disease, their parents and siblings: an overview of the research evidence base. *Child Care Health Dev.* 2006;32:19–31.

22. Rutter M. Psychosocial resilience and protective mechanisms. In: Rolf J, Masten D, Cicchetti K, et al., eds. *Risk and Protective Factors in the Development of Psychopathology.* Cambridge: Cambridge University Press; 1990:181–214.

23. Thompson RJ, Gustafson KE, Hamlett KW, Spock A. Stress, coping and family functioning in the psychological adjustment of mothers of children and adolescents with cystic fibrosis. *J Pediatr Psychol.* 1992;17:573–585.

24. DeMaso DR, Campis LK, Wypij D, Bertram S, Lipshita M, Freed M. The impact of maternal perceptions and medical severity on the adjustment of children with congenital heart disease. *J Pediatr Psychol.* 1991;16:137–149.

25. Hovels-Gurich HH, Konrad K, Wiesner M, et al. Long term behavioural outcomes after neonatal arterial switch operation for transposition of the great arteries. *Arch Dis Child.* 2002;87:506–510.

26. Wright M, Nolan T. Impact of cyanotic heart disease on school performance. *Arch Dis Child.* 1994;71:64–70.

27. Van Rijen EH, Utens EM. Psychological aspects of congenital heart disease in children. In: Wyszynski D, Correa-Villasenor A, Graham T, eds. *Congenital Heart Defects, From Origin to Treatment.* Oxford University Press; 2010.

28. Achenbach TM, Rescorla LA. *Manual for the ASEBA School-Age Forms and Profiles.* Burlington, VT: University of Vermont, Research Center for Children, Youth and Families; 2001.

29. Casey FA, Sykes DH, Craig BG, Power R, Mulholland HC. Behavioural adjustment of children with surgically palliated complex congenital heart disease. *J Pediatr Psychol.* 1996;21:335–352.

30. McCusker CG, Doherty NN, Molloy B, et al. Determinants of neuropsychological and behavioural outcomes in early childhood survivors of congenital heart disease. *Arch Dis Child.* 2007;92:137–141.

31. McCusker CG, Armstrong MP, Mullen M, Doherty NN, Casey F. A sibling-controlled prospective study of outcomes at home and school in children with severe congenital heart disease. *Cardiol Young.* 2013;23(4):507–516.

32. Pinquart M, Shen Y. Behavior problems in children and adolescents with chronic physical illness: a meta analysis. *J Pediatr Psychol.* 2011;36:1003–1016.

33. Dixon Woods M, Cavers D, Agarwal S, et al. Conducting a critical interpretive synthesis of the literature on access to healthcare by vulnerable groups. *BMC Med Res Methodol.* 2006. http://dx.doi.org/10.1186/1471-2288-6-35.

34. Drotar D. *Psychological Interventions in Childhood Chronic Illness.* Washington, DC: American Psychology Association; 2006.

35. Casey FA, Stewart M, McCusker CG, et al. Examination of the physical and psychosocial determinants of health behaviour in 4–5 year old children with congenital heart disease. *Cardiol Young.* 2010;20:532–537.

III. PSYCHOLOGICAL PROFILES AND PROCESSES

36. Bayley N. *Bayley Scales of Infant Development*. 2nd ed. San Antonio, TX: Harcourt Brace & Company; 1991.
37. Derogatis IR. *Brief Symptom Inventory: Administration, Scoring and Procedures Manual*. Minneapolis, MN, USA: National Computer Systems; 1993.
38. Moos RH, Moos BS. *Family Environment Scale*. New York: McGraw-Hill; 1991.
39. Bellinger DC, Newburger J, Wypij D, Kuban K, DuPlessis A, Rappaport L. Behavior at eight years in children with surgically corrected transposition the Boston Circulatory Arrest Trial. *Cardiol Young*. 2009;19:86–97.
40. Bellinger DC. Are children with congenital cardiac malformations at increased risk of deficits in social cognition. *Cardiol Young*. 2008;18:309.
41. Utens EM, Verhulst FC, Meijboom FJ, et al. Behavioural and emotional problems in children and adolescents with congenital heart disease. *Psychol Med*. 1993;23:415–424.
42. Miatton M, DeWolf D, Francois K, Thiery E, Vingerhoets G. Behaviour and self-perception in children with a surgically corrected congenital heart disease. *J Dev Behav Pediatr*. 2007;28:294–301.
43. Fredriksen PM, Mengshoel AM, Frydenlund A, Sorbye O, Thaulow E. Follow-up in patients with congenital heart disease more complex than haemodynamic assessment. *Cardiol Young*. 2004;14:373–379.
44. Davidson J, Gringras P, Fairhurst C, Simpson J. Physical and neurodevelopmental outcomes in children with single-ventricle circulation. *Arch Dis Child*. 2014. http://dx.doi.org/10.1136/archdischild-2014-306449.
45. Freitas IR, Castro M, Sarmento SL, et al. A cohort study on psychosocial adjustment and psychopathology in adolescents and young adults with congenital heart disease. *BMJ Open*. 2013. http://dx.doi.org/10.1136/bmjopen-2012-001138.
46. Van Rijin EH, Utens EM, Roos-Hesselink JW, et al. Longitudinal development of psychopathology in an adult congenital heart disease cohort. *Int J Cardiol*. 2005;99:292–298.
47. Spijkerboer AW, De Koning WB, Duivenvoorden HJ, et al. Medical predictors for long-term behavioural and emotional outcomes in children and adolescents after invasive treatment of congenital heart disease. *J Pediatr Surg*. 2010;45:2146–2153.
48. Spijkerboer AW, Utens EM, Bogers AJ, Helbing WA, Verhulst FC. A historical comparison of long-term behavioral and emotional outcomes in children and adolescents after invasive treatment for congenital heart disease. *J Pediatr Surg*. 2008;43:534–539.
49. Marino BS, Lipkin PH, Newburger JW, et al. Neurodevelopmental outcomes in children with congenital heart disease: evaluation and management: a scientific statement from the American Heart Association. *Circulation*. August 28, 2012;126(9):1143–1172.
50. Le Roy S, Elixson EM, O'Brien P, et al. Recommendations for preparing children and adolescents for invasive cardiac procedures: a statement from the American Heart Association Pediatric Nursing Subcommittee of the Council on Cardiovascular Nursing in collaboration with the Council on Cardiolvascular Diseases of the Young. *Circulation*. 2003;108:2550–2564.

A Family Affair

N. Doherty

Western Health and Social Care Trust, Derry, Northern Ireland

E. Utens

Erasmus Medical Center, Sophia Children's Hospital, Rotterdam, The Netherlands

OUTLINE

THE IMPORTANCE OF FAMILY FUNCTIONING FOR THE CHILD WITH CONGENITAL HEART DISEASE

No child exists in isolation. Surrounding each child, and influencing each child, is the family. It has long been recognized in the child psychology literature that families are central to the subsequent development of the child, but this recognition has found greater public voice and has become influential in health-care policy.[1–3] Indeed in the previous chapters we have noted how family factors such as maternal mental health, illness-related worry, family cohesion, and parenting styles are often more significant than disease or surgical factors in predicting long-term outcomes for the child with congenital heart disease (CHD).

The importance of family factors has been demonstrated by other research groups as well, with mental health, coping styles and adjustment in parents shown to accentuate risk, or confer protective benefits, on the developing child with CHD—and indeed the extended family.[4–8] This includes longitudinal demonstration that levels of distress in parents of children with CHD in infancy predicts child adjustment years later.[8] Mussato offers a multidirectional model to help understand how such influences might operate,[9] and two specific examples are considered in the following sections.

The task of establishing a secure attachment was noted, for example, in the chapter "Is There a Behavioral Phenotype for Children with Congenital Heart Disease?," to be significantly challenged by congenital heart disease. Thus, the fragile and sick infant, often too breathless to feed easily, subjected to multiple invasive procedures and often separated from the primary caregiver, will significantly challenge the caretaking and interactional behaviors fundamental to the development of a secure attachment. The most well adjusted of parents will find this difficult to manage and contain. Indeed studies have shown that children with CHD are less likely to be securely attached than healthy infants, or indeed infants born with cystic fibrosis,[10] and that the severity of the disease exerts a negative influence here.[11] Consider, however, the impact of such stressors on a vulnerable mother—perhaps with low self-esteem, poor parenting capacities, and mental health difficulties. As the authors have noted, the mother's own attachment style and personal disposition impact independently on this process either amplifying or reducing the risk.[10,11] The *toxic* effect of depressed or disturbed *state of mind* in the mother can become internalized in the infant and exacerbate current and indeed later negative effects.[12]

Another developmental task for children which can be compromised by CHD includes participation in sports and exercise. Physical factors, but especially *expectations* of reduced exercise capacity and tolerance (see chapter: Is There a Behavioral Phenotype for Children with Congenital Heart Disease?), often erroneous, can challenge this. Interventions in this domain are further considered in chapter "Healthy Teenagers and Adults: An Activity Intervention," but a study conducted in 2015 in our center in the Netherlands highlights again the impact of parental mental health on this aspect of child adjustment.[13] The study consisted of a multicenter, prospective, randomized controlled, intervention study into the effect of a 12-week standardized exercise training program in a cohort of children and adolescents, aged 10–15 years, with either tetralogy of Fallot (ToF) or a Fontan circulation. Results suggested a moderating influence of parental mental health: compared with controls, adolescents in the exercise group who had anxious parents, showed a significant decrease in social functioning after the exercise program. This suggested that poor parental mental health may hamper or even undermine expected improvements of exercise training. The authors suggested that this may have been through modeling elevated levels of anxiety and worry as associated with exercise participation. Clearly if such processes were occurring in a study trial, these may be expected to be more pronounced in everyday (uncontrolled) exercise scenarios for the child and adolescent with CHD. Findings demonstrate that interventions targeted at the child, but neglecting the important systemic influence of the parents, may be undermined in efficacy and impact.

PARENTS OF CHILDREN WITH CHD

Given the importance of family factors in mediating outcomes for children with CHD, it is important to consider how these key participants in the child's world may be affected by the disease. At the outset it is important to note that, in many ways, findings here mirror findings with the children themselves (see chapter: Is There a Behavioral Phenotype for Children with Congenital Heart Disease?), in that this is not always a negative picture. For example, in our Dutch center two studies focused on long-term parental mental health outcomes.[14,15] The children had different congenital heart defects. Interestingly, results showed that in the long term (at least 7 years after cardiac surgery/catheter intervention in the child with CHD), parents actually reported better mental health compared to normative data. These parents reported less distress, less somatic symptoms, less anxiety, sleeplessness, or serious depression than parents from the normative reference groups. These "better than normal," very favorable, results were explained by the parents themselves. During semi-structured interviews, parents often suggested that the stressful experiences related to the CHD diagnosis and surgery had made them stronger. They reported that they worried less about the *futilities* in life. This pattern can be called *posttraumatic growth*.

In this study more favorable outcomes on coping were also found, compared to normative data. These parents less often expressed negative emotions (anger, annoyance) and less often needed to use reassuring thoughts (ones which are used in times of stress and anxiety). It has to be taken into account that in these studies most patients did not undergo rehospitalizations. Moreover, the long-term medical course was stable for the majority of them. However, other centers have found normal levels of psychological functioning in parents of children following cardiac surgery, especially as time elapses since the surgery and a "normal" life can become reinstated.[16]

Nevertheless, such positive outcomes have not always been found and a review of research in this area in 2012 by Soulvie et al.[17] highlighted elevated levels of parenting stress and distress in families of children with CHD, and especially through health-care experiences. These authors highlighted the need for support to be provided to parents especially to enhance the quality of the parent–child interaction which they argued would impact on later child outcomes. Specific findings, related to elevated maternal mental health difficulties, have been observed and noted to impact on subsequent "internalizing" behavior problems in the child with CHD.[18]

The importance of parental distress and coping in infancy has been noted above to be vital for much later child adjustment, and work at our Belfast center examined psychological functioning and coping styles in both mothers and fathers in the early months of life of their infant born with severe CHD.[19] The results indicated that one-third of mothers and almost one-fifth of fathers experienced levels of psychological disturbance which reached clinical "caseness" on the *Brief Symptom Inventory*[20] in the months following the birth of a child with CHD (see Table 7.1 below). These figures are significant and in terms of paternal mental health (rarely looked at in the literature) are likely to represent an underestimation as it has been suggested that fathers will seek to appear "strong" on preliminary questioning and in the absence of more sensitive and detailed interviewing about their experience.[21,22] As with the Visconti study noted above,[8] when we followed up the infants of the parents included in this study 7 years later, subsequent behavioral adjustment in the children was found to be predicted by the levels of psychological distress manifested by their parents when they were infants in the study described here.[23]

As is discussed further in the following paragraphs, coping styles were significant in mediating levels of distress in both parents. However, it is worth highlighting that despite the apparently higher levels of distress reported by mothers, the authors noted concerns that some of the more adaptive coping styles (seeking instrumental and emotional support and talking more) were significantly more preponderant in mothers compared to fathers; fathers, on the other hand, showed significantly greater use of maladaptive coping styles (eg, disengagement through use of alcohol).

Qualitative research has perhaps been required to extend our understanding of fathers' experiences. Themes labeled as *relinquishing and*

TABLE 7.1 Summary Table for Parental Scores on the Brief Symptom Index Detailing Mean Composite Score, Standard Deviation, and Percentage of Scores Within Levels for Clinical Caseness

	Mean Composite Score (GSI)	Standard Deviation	Clinical Caseness (%)
Mothers	0.60	0.60	33
Fathers	0.42	0.62	18

reclaiming control and *living in the shadow of the illness* were explicated in one key study which suggested that fathers were struggling in perhaps quite unique ways.[22] The authors described a struggle with managing their identity as family "protector" in the face of CHD, in assuming a lead role on pragmatic initiatives at the expense of emotional ventilation and with playing a *support act* in hospital interfaces. We do need to better understand fathers' perspectives and experiences here as they may not always fit familiar models of stress and coping in this domain (due to their absence from research which has informed these narratives). This is important as the rare research in the pediatric literature, which has focused explicitly on their involvement in the chronic disease management of their children, has highlighted that the greater such active involvement is, the better the outcomes for maternal, marital, and overall family functioning.[24]

Finally, although it has received little attention in the CHD literature to date, it is important to look at the experience of siblings and indeed on bidirectional influences with the child with CHD. The pediatric psychology literature on siblings in general is small and, as with parents, findings can appear contradictory. Some have found no discernible negative impact,[25,26] or indeed benefits in terms of increased prosocial behavior.[27] However, meta-analysis in this area does point toward increased risk for some degree of maladjustment.[28] Preliminary work from our own center in Belfast has again used qualitative methods (Interpretive Phenomenological Analysis, IPA[29]) to better understand siblings' experience. Siblings in this study scored in the "normal" range on a standardized test of behavioral adjustment,[30] and would thus appear well adjusted on the basis of traditional quantitative analyses. Importantly, however, analysis of the narratives of their experience suggested a much more complex process. Themes related to *a family fractured*, *defending against distress*, *the struggle for balance*, and *looking forward but thinking back*, highlighted that these young people had been *walking the line* between processes which could at once exacerbate risk or promote resilience. The authors concluded that adjustment in siblings was not an all or nothing "outcome," but rather was a fluid process, the outcome of which would depend on other situational factors in their personal and family lives. Nevertheless, sibling CHD clearly represented an important factor in their lives and precipitated processes relating to managing distress, dissonance, and family disruptions which are significant in the etiology of maladjustment in children per se. Of interest, the study itself highlighted the therapeutic benefits reported by the siblings who participated in the study. This appeared to be because it offered often the first opportunity for them to "construct meaning" surrounding their experiences, a process which would become an important element of our interventional work, which is discussed in later chapters.

III. PSYCHOLOGICAL PROFILES AND PROCESSES

FACTORS PREDICTING OUTCOMES FOR FAMILIES

A central argument of this book has been that parental and family functioning is central to understanding outcomes for children with CHD. However, in this chapter we have highlighted that these parents and families themselves are vulnerable to maladjustment. This points toward providing family-based interventions in CHD which is being increasingly recognized as important in pediatric psychology more generally.[31] However, to target such interventions most effectively we need to understand the risk and protective factors for the parents and families themselves. There are interesting parallels here to findings with the children themselves, as discussed in chapter "Is There a Behavioral Phenotype for Children with Congenital Heart Disease?".

While Majmnemer et al.[32] reported, in a small study, that arterial oxygen saturation, with levels below 85% prior to surgery, predicted parental stress in children with various types of CHD, more often no clear relationship between disease severity and parental outcomes has been demonstrated as might be expected. As highlighted by Dulfer, Helbing, and Utens in their review of studies into parental coping following CHD,[33] the majority of studies reported in 2003, 2004, and 2009 have not shown any distinct association between disease severity and parental stress or coping. Mussatto reported, for example, that disease severity was not a reliable predictor for long-term parental psychosocial outcomes,[9] and others have generally shown a dissociation between disease severity and levels of parental stress.[16,34,35]

In chapter "Historical Perspectives in Pediatric Psychology and Congenital Heart Disease" it was noted that unidirectional models (disease parameters leading to maladjustment) were not likely to fully capture etiological mechanisms and that systemic models were required to understand outcomes for children with chronic illnesses and their families. Bronfenbrenner's ecological model of development[36] highlighted the importance of formal and informal systems of influence around the individual. The importance of social networks in mediating the posttraumatic stress experience of parents of children with chronic illness was well expounded early on by Kazak et al.[37] Similar findings have been reported in 2002, 2006, and 2014 in the population of parents whose children have congenital heart disease.[38-40] Tak and McCubbin in 2007, for example, found in 92 families of children newly diagnosed with CHD that perceived social support acted as a resiliency factor between family stress and parental coping.[39] Lawoko and Soares studied a large sample of 632 parents of children with CHD (aged 0–20 years) and also found that decreasing availability of social support, over a 1-year period, was related to increased risk for psychological morbidity (eg, anxiety and depression).[40] The fact that fathers tend to use social support less often than mothers[15,19] should make us particularly concerned about how they are coping despite the apparently reduced levels of psychological morbidity.

TABLE 7.2   Predictors of Maternal Mental Health ($R2 = 0.44$, $F = 14.6$, $P < 0.001$)

Variable	Beta	t	p
Coping—behavioral disengagement	0.355	3.25	0.002
Understanding of diagnosis	−0.28	−2.56	0.014
Family—cohesion	−0.33	−2.95	0.005

However, we need to think more precisely about interpersonal and indeed intrapersonal resources around families and, in particular, we need to identify those factors which we might be able to do something about. Lawoko and Soares,[40] for example, highlighted that in addition to perceptions of social support, specific perceptions related to the "burden of care" and quality of health-care interactions also contributed to variance on levels of parental distress. Our own work at the Belfast center reported in earlier paragraphs took this further.[19] Again we did not find disease, comorbid diagnoses, or surgical factors to be particularly relevant in understanding the levels of parental distress reported in the previous paragraphs. Indeed in this study socioeconomic status and perceptions of social support were not the most important predictors of outcome. Rather, for mothers, intrapersonal coping skills (especially degree of behavioral engagement in caretaking behaviors), degree of family cohesiveness, and the degree of understanding of their child's diagnosis predicted levels of psychological distress (see Table 7.2). A similar pattern was found in fathers. These findings highlight the importance of the medical system surrounding the child and family and in particular that effective communication and engagement of parents in the care of their child is instrumental in promoting parental adaptation. These are processes open to intervention. Moreover, findings highlight that personal and family attributes related to coping skills and cohesiveness are also protective. Again, while we may not be able to alter the fact of the congenital heart defect, these psychosocial attributes may indeed be something which we can bolster and shape.

SUMMARY: LEARNING POINTS AND MOVING FORWARD IN CLINICAL PRACTICE

In conclusion, while distress levels can normalize, or may even become lower compared to normative data through a process akin to posttraumatic growth, parents of children with CHD are generally at elevated risk for psychological difficulties at key points in the course of their child's illness and treatment. Importantly, how well parents can adjust to living with their child with CHD influences the developmental outcomes of the child, and thus we suggest that interventions should be family based.

At the very least it is important that teams working with children with CHD and their families are multidisciplinary with psychological representation so that professional psychological help, advice, and support are accessible to both staff and families. This may happen by provision of some general material to families and increasing knowledge of psychological information at ward level, or it may be more specific input for psychological support. We do not think that every family needs extended psychological intervention. Such may only be required if there are preexisting risk factors such as mental health or family difficulties. However, there is also a case for secondary prevention initiatives in this high-risk population.

Jackson et al. in their 2015 review of parental coping in CHD have called for this to be holistic and aimed at bolstering the parent–child relationship.[41] Our own work has demonstrated, for example, that showing mothers how to engage in a simple infant massage intervention when their child is in hospital (see Picture 7.1),[42] resulted in increased reports of attachment

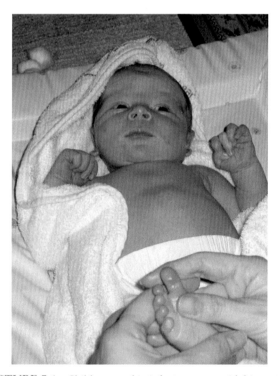

PICTURE 7.1   Child engaged in infant massage with his mother.

and reduced anxiety and stress levels surrounding their child's hospital stay. As noted above, we have indicated how opportunities for siblings to "construct meaning" surrounding their family experiences can be beneficial, and De Maso and colleagues, some time ago, showed that similar narrative therapy interventions online can help mothers cope better with their child's CHD.[43] However, it will be in the fourth part of this book where we look at how holistic family interventions, of a relatively brief nature, can be provided systematically and at key developmental transitions to improve outcomes for both the parents and their children.

References

1. Wachs TD, Black MM, Engle PL. Maternal depression: a global threat to children's health, development, and behavior and to human rights. *Child Dev Perspect*. 2009;3:51–59.
2. Chang JJ, Halpern CT, Kaufman JS. Maternal depressive symptoms, father's involvement, and the trajectories of child problem behaviors in a US national sample. *Archives Pediatr Adolesc Med*. 2007;161:697–703.
3. Marryat L, Martin C. *Growing up in Scotland: maternal mental health and its impact on child behaviour and development*. Edinburgh: Scottish Government; 2010.
4. Carey LK, Nicholson BC, Fox RA. Maternal factors related to parenting young children with congenital heart disease. *J Ped Nurs*. 2002;17:174–183.
5. Hulser K, Dubowy KO, Knobl H, Meyer H, Scholmerich A. Developmental outcome and psychosocial adjustment in children after surgery for congenital heart disease during infancy. *J Reprod Infant Psychol*. 2007;25:139–151.
6. Wallander JL, Varni JW. Adjustment in children with chronic physical disorders: programmatic research on a disability–stress–coping model. In: La Greca AM, Siegel L, Wallander JL, Walker CE, eds. *Stress and coping in child health*. New York: Guilford Press; 1992:279–298.
7. Goldberg S, Janus M, Washington J, Simmons RJ, MacLusky I, Fowler RS. Prediction of preschool behavioural problems in healthy and pediatric samples. *J Dev Beh Ped*. 1997;18:304–313.
8. Visconti KJ, Saudino KJ, Rappaport LA, Newburger J, Bellinger DC. Influence of parental stress and social support on the behavioral adjustment of children with transposition of the great arteries. *J Dev Beh Pediatr*. 2002;23:314–321.
9. Mussatto K. Adaptation of the child and family to a life with chronic illness. *Cardiol Young*. 2006;16:110–116.
10. Goldberg S, Simmons RJ, Newman J, Campbell K, Fowler RS. Congenital heart disease, parental stress and infant–mother relationship. *J Pediatr*. 1991;119:661–666.
11. Berant E, Mikulincer M, Florian V. The association of mothers' attachment style and their psychological reactions to the diagnosis of infant's congenital heart disease. *J Soc Clin Psychol*. 2001;20:208–232.
12. Tronick EZ, Weinberg MK. Depressed mothers and infants: failure to form dyadic states of consciousness. In: Murray L, Cooper PJ, eds. *Postpartum depression and child development*. New York: Guilford Press; 1997:54–81.
13. Dulfer K, Duppen N, Van Dijk AP, Kuipers IM, Van Domburg RT, Verhulst FC, et al. Parental mental health moderates the efficacy of exercise training on health-related quality of life in adolescents with congenital heart disease. *Pediatr Cardiol*. January 2015;36(1):33–40.
14. Utens EM, Versluis-Den Beiman HJ, Witsenburg M, Bogers AJ, Hess J, Verhulst FC. Does age at the time of elective cardiac surgery or catheter intervention in children influence the longitudinal development of psychological distress and styles of coping of parents? *Cardiol Young*. 2002;12(6):524–530.

15. Spijkerboer AW, Helbing WA, Bogers AJ, Van Domburg RT, Verhulst FC, Utens EM. Long-term psychological distress, and styles of coping, in parents of children and adolescents who underwent invasive treatment for congenital heart disease. *Cardiol Young.* 2007;17:638–645.
16. Wray J, Sensky T. Psychological functioning in parents of children undergoing elective cardiac surgery. *Cardiol Young.* 2004;14(2):131–139.
17. Soulvie MA, Desai PP, Parker White C, Sullivan BN. Psychological distress experienced by parents of young children with congenital heart defects: a comprehensive review of literature. *J Soc Serv Res.* 2012;38(4):484–502.
18. Landolt MA, Ystrom E, Stene-Larsen H, Holstrom H, Volrath ME. Exploring causal pathways of child behaviour and maternal mental health in families with a child with congenital heart disease: a longitudinal study. *Psychol Med.* 2014;44:3421–3433.
19. Doherty N, McCusker CG, Molloy B, Mulholland C, Rooney N, Craig B, et al. Predictors of psychological functioning in mothers and fathers of infants born with severe congenital heart disease. *J Reprod Infant Psychol.* 2009;27(4):390–400.
20. Deragotis LR. *BSI: brief symptom inventory. Administration, scoring and procedures manual.* Minneapolis, MN: National Computer Systems; 1993.
21. Phares V, Lopez E, Fields S, Kamoukos D, Duhig A. Are fathers involved in paediatric psychology research and treatment? *J Ped Psychol.* 2005;30:631–643.
22. Chesler M, Parry C. Gender roles and/or styles in crisis: an integrative analysis of the experiences of fathers of children with cancer. *Qual Health Res.* 2001;11:363–384.
23. McCusker CG, Armstrong MP, Mullen M, Doherty NN, Casey FA. A sibling-controlled, prospective study of outcomes at home and school in children with severe congenital heart disease. *Cardiol Young.* 2013;23:507–516.
24. Gavin L, Wysocki T. Associations of paternal involvement in disease management with maternal and family outcomes in families with children with chronic illness. *J Ped Psychol.* 2006;31(5):481–489.
25. Sahler OJZ, Roghmann KJ, Carpenter PJ, Mulhern RK, Dolgin MJ, Sargent JR. Sibling adaption to childhood cancer collaborative study: prevalence of sibling distress and definition of adaption levels. *Dev Beh Pediatr.* 1994;15:353–366.
26. Fleary SA, Heffer RW. Impact of growing up with a chronically ill sibling on well siblings' late adolescent functioning. *Fam Med.* 2013:Article ID 737356. http://dx.doi.org/10.5402/2013/737356.
27. Houtzager BA, Grottenhuis MA, Caron HN, Last B. Quality of life and psychological adaption in siblings of pediatric cancer patients, 2 year after diagnosis. *Psycho-Oncol.* 2004;13(8):499–511.
28. Verames I, Van Susante AM, Hedwig JA. Psychological functioning of siblings in families of children with chronic health conditions: a meta-analysis. *J Ped Psychol.* 2012;37(2):166–182.
29. Smith JA, Flowers P, Larkin M. *Interpretative phenomenological analysis.* London: Sage; 2009.
30. Kennedy L, McCusker CG, Russo K. *Walking the line between risk and resilience: the experience of young people who have a brother or sister living with congenital heart disease.* Thesis submitted in partial fulfilment of the requirement of Doctorate in Clinical Psychology. Queens University Belfast; 2014.
31. Law EF, Fisher E, Fales J, Noel M, Eccleston C. Systematic review and meta analysis of parent and family-based interventions for children and adolescents with chronic medical conditions. *J Ped Psychol.* 2014;39:866–886.
32. Majnemer A, Limperopoulos C, Shevell M, Rohlicek C, Rosenblatt B, Tchervenkov C. Health and well-being of children with congenital cardiac malformations, and their families, following open-heart surgery. *Cardiol Young.* 2006;16(2):157–164.
33. Dulfer K, Helbing WA, Utens EM. Coping in parents of children with congenital heart disease. In: Molinelli B, Grimaldo V, eds. *Handbook of the psychology of coping: new research.* Hauppauge, NY: Nova Science; 2012:307–320.

34. Uzark K, Jones K. Parenting stress and children with heart disease. *J Pediatr Health Care.* 2003;17(4):163–168.

35. Vrijmoet-Wiersma CM, Ottenkamp J, van Roozendaal M, Grootenhuis MA, Koopman HM. A multicentric study of disease-related stress, and perceived vulnerability, in parents of children with congenital cardiac disease. *Cardiol Young.* 2009;19(6):608–614.

36. Bronfenbrenner U. *The ecology of human development: experiments by nature and design.* Cambridge: Harvard University Press; 1979.

37. Kazak AE, Barakat LP, Meeske K, Christakis D, Meadows AT, Casey R, et al. Posttraumatic stress, family functioning and social support in survivors of childhood leukaemia and their mothers and fathers. *J Consult Clin Psychol.* 1997;1:120–129.

38. Werner H, Latal B, Valsangiacomo Buechal E, Beck I, Landolt MA. The impact of an infant's severe congenital heart disease on the family: a prospective cohort study. *Congen Heart Dis.* 2014;9:203–210.

39. Tak YR, McCubbin M. Family stress, perceived social support and coping following the diagnosis of a child's congenital heart disease. *J Adv Nurs.* 2002;39:190–198.

40. Lawoko S, Soares JJ. Psychosocial morbidity among parents of children with congenital heart disease: a prospective longitudinal study. *Heart Lung.* 2006;35(5):301–314.

41. Jackson AC, Frydenberg E, Liang RP, Higgins RO, Mirphy BM. Familial impact and coping with child heart disease: a systematic review. *Ped Cardiol.* 2015;36:695–712.

42. Tierney N, Doherty N, Casey F, Craig B, Sands A. Impact of an infant massage programme on maternal mental health and attachment. *J Reprod Infant Psychol.* 2008;26(3): 256–271.

43. De Maso DR, Gonzalez-Heydrich J, Erickson JD, Grimes VP, Strohecker C. The experience journal: a computer-based intervention for families facing congenital heart disease. *J Am Acad Child Adoles Psychiat.* 2000;39:727–734.

8

The Adult With Congenital Heart Disease

D. Katz

York University, Toronto, ON, Canada

M. Chaparro

University of Toronto, Toronto, ON, Canada

A.H. Kovacs

University Health Network, Toronto, ON, Canada; University of Toronto, Toronto, ON, Canada

OUTLINE

INTRODUCTION

Beginning in the mid-20th century, there have been vast changes in the diagnosis and treatment of congenital heart disease (CHD). With improvements in care, the median age of individuals with CHD has increased and the number of adults living with CHD continues to rise.[1] As of 2010, it was estimated that two-thirds of those with CHD were adults, and in Canada alone there were estimated to be over 166,000 adults living with CHD.[2] Despite improvements in pediatric care, CHD is rarely "cured."[1] Rather, adults with CHD, particularly those with defects of moderate or great complexity, often develop complications such as arrhythmias and heart failure, undergo additional cardiac surgeries and interventions, and remain at risk of premature death.[1,3–6] Lifelong surveillance by CHD specialists is therefore recommended.[7] This emerging population of adults with CHD thus constitutes a group with unique medical as well as psychosocial concerns.

The psychosocial experiences of adults with CHD differ from those of adults with other forms of acquired heart disease (such as atherosclerosis) that commonly present in older adulthood. Moderate or complex CHD is typically diagnosed at birth or infancy and is being increasingly diagnosed in the prenatal period. Adults with CHD thus face the challenge of coping with a lifelong medical condition, thereby requiring individualized education and support from their health-care teams. Many individuals describe growing up with the sense of "feeling different" from others and report difficulties with social functioning.[8–11] These early life experiences can have later reverberations in adulthood, leading to an elevated risk for symptoms of anxiety and depression.[12]

Although distinct from adults with acquired heart disease, individuals with CHD do not themselves represent a homogeneous patient cohort. Adults with CHD differ from one another in terms of original CHD diagnosis, disease complexity, treatment history, and both cardiac and noncardiac comorbidities and sequelae. For some patients, CHD may be experienced as a chronic illness that regularly impedes their daily functioning. For others, CHD may have minimal impact on their lives or identities. Certain cardiac symptoms and interventions may also contribute to varying experiences among patients. For example, individuals prone to ventricular arrhythmias and at risk for sudden cardiac death may receive an implantable cardioverter defibrillator (ICD).[13] Living with an ICD, in turn, is associated with several specific psychosocial challenges, such as anxiety regarding future device shocks.[14,15] Advanced heart failure, a common cause of deteriorating health and mortality among adults with CHD,[3,16] is associated with increased prevalence of depression.[17,18] In summary, one cannot make assumptions about the psychosocial well-being of any one individual with CHD.

As the number of adults with CHD continues to grow, it becomes increasingly important to explore and describe the psychosocial pheno-type of this distinct yet diverse population. An improved conceptualiza-tion of the psychological concerns and challenges faced by adults with CHD could lead to the provision of appropriate support and interventions to most effectively meet their needs.

PSYCHOSOCIAL FUNCTIONING

Growing up with a serious and often unpredictable medical condition poses unique psychosocial challenges for the lives of adults with CHD. In addition to medical care-related challenges (eg, treatment decision making, adjustment to implanted cardiac devices, surgical preparation, and adjustment to declining physical health status), impact may also be observed in social, educational, and emotional domains.[19] As a group, adults with CHD face an elevated risk of neurocognitive problems, surgi-cal scarring that may contribute to body image concerns, the possibility of a sudden decline in physical functioning, unique family planning con-siderations, identity and autonomy formation needs, and a shortened life expectancy.[9,20–24] Understandably, coping with these challenges may put adults with CHD at risk for elevated psychosocial distress, particularly anxiety and depression.

Psychological Distress

A systematic review in 2013 concluded that although the prevalence of anxiety and depression in adults with CHD is comparable to the preva-lence found in adults with acquired heart disease, it is higher than that which has been observed in the general population without heart dis-ease.[12] However, due to variation in the assessment of psychosocial dis-tress (eg, psycho-diagnostic clinical interviews vs. self-report inventories) and apparent international differences in the quality of life of adults with CHD, research has produced heterogeneous findings.[25]

Three North American studies have documented elevated psycho-logical distress in adults with CHD using clinical interview methodology. In one study examining the prevalence of depression and anxiety, 14% met criteria for major depressive disorder and 17% for panic disorder.[10] A significant proportion also reported subclinical symptomatology for other anxiety, mood, or adjustment disorders. Another study assessed the prevalence of depressive and anxiety disorders in "well-adjusted" patients using a semi-structured standardized psychiatric interview and a validated self-report questionnaire.[26] Clinical interviews revealed that

27% met criteria for a depressive episode and 9% met criteria for generalized anxiety disorder. A self-report measure detected that 27% of participants scored above a clinical cutoff for either depression or anxiety. In a third study, the prevalence of mood and anxiety disorders among 58 adults with CHD recruited from one of two outpatient clinics was evaluated using a standardized clinical interview.[27] The authors found that 50% of the sample met diagnostic criteria for a lifetime mood or anxiety disorder (33% mood disorder and 26% anxiety disorder) and that 29% met diagnostic criteria at the time of the assessment. In a larger sample of 280 patients within the same study, survey responses revealed that 12% endorsed symptoms of moderate to severe depression and 34% reported symptoms of elevated anxiety. In summary, results of North American studies that employed structured diagnostic interviews suggest that about one in three adults with CHD experiences clinically significant depression and/or anxiety.

There are also studies from outside of North America that have observed similar findings. For example, the prevalence of depression was assessed in a sample of Czech adults with CHD with persistent cyanosis using a standardized self-report questionnaire.[28] Thirty-four percent of patients in this sample received scores consistent with depression. In a 2014 study from Greece, 28% of patients endorsed elevated symptoms of depression.[25]

Findings regarding the incidence of psychological distress, however, are certainly not unequivocal as some other studies have documented little or no differences between adults with CHD and their healthy peers, particularly in Western European samples.[29–31] Studies from the Netherlands have found that adults with CHD are slightly better than healthy controls with regard to hostility and neuroticism.[30,31] Although there is indeed variability across studies, the increased prevalence of anxiety and depression observed by many researchers suggests that this area remains an important focus of ongoing clinical attention and research.

Correlates of Psychological Distress

Given the heterogeneity of the CHD population with regard to medical, psychosocial, and demographic profiles, an exploration of potential correlates of psychological distress is warranted.

Medical variables. With few exceptions, the absence of an association between medical-related variables (eg, CHD diagnosis, disease severity, and age at first surgical procedure) and psychological distress has been supported by several studies.[26–28,30,32] It should be noted, however, that there are studies that have shown that severe CHD and surgical repair have been associated with poorer psychosocial adjustment.[33,34] Moreover, adults with heart disease who experience ICD implantation, heart failure, or heart and/or lung transplantation tend to face less favorable psychological outcomes.[14,15,35,36]

Psychosocial variables. The specific psychosocial context in which adults with CHD face their illness also play a role in their levels of psychological distress. The following have been associated with poorer psychological functioning in samples of adults with CHD: limited social support, loneliness, low independence, maternal overprotection, poor academic performance, reduced physical competence, low social problem-solving skills, imposed limits, social anxiety, lower self-esteem, and negative thinking.[17,27,30,33,37–39]

Demographic variables. Studies that have examined whether psychological functioning among adults with CHD differs as a function of demographic variables has yielded mixed results. A 2013 study reported a higher prevalence of psychopathology in females than males,[33] although other studies have not detected a link between sex and levels of anxiety and depression.[26,27,31,40] There are also conflicting outcomes with respect to age. Some studies have found that age is not linked with psychological distress,[26,40] whereas others have found that older participants tend to have worse psychological outcomes.[28] Moreover, although one study observed no differences in psychological outcomes based on living arrangements, marital status, education, or employment status,[26] other studies have found that better psychological functioning was associated with being in a stable romantic relationship, having a higher level of education, and being employed.[28,40] In summary, due to the conflicting nature of results, it is not possible to draw any definitive conclusions regarding the influence of socio-demographic variables on the psychological outcomes of adults with CHD.

Unique Psychosocial Challenges

It is important that we do not restrict our exploration of the psychological needs of adults with CHD to traditional mental health concerns, namely, anxiety and depression. Rather, consistent with the same developmental framework outlined in chapter "Is There a Behavioral Phenotype for Children With Congenital Heart Disease?," it may be more apposite to consider the specific psychosocial challenges of adolescence, adulthood, and older adulthood which CHD challenge and which may amplify risk for formal psychiatric diagnoses depending on other contextual factors.

Having CHD often results in additional concerns related to the self and others. A qualitative study explored some of these concerns as well as the motivations for seeking mental health treatment.[9] Intrapersonal challenges included body image concerns related to surgical scarring and tolerating uncertainty about the long-term health expectations. Interpersonal challenges included growing up with a sense of "feeling different," experiencing social isolation, dealing with conflicting social expectations, and having difficulties talking to others about their illness. Similar concerns and dilemmas have been identified in other studies, including difficulty balancing independence with dependence and feeling integrated within a social environment while occasionally facing limitations in doing so.[21,41,42] Other documented challenges include heart-focused anxiety and somatic complaints.[19] Concerns about sexual intimacy, fertility, contraception, and pregnancy are also common in this population.[22] There are two specific psychosocial challenges that warrant additional attention, namely, the transition from pediatric to adult care and the navigation of end-of-life concerns.

Transition

The transition from pediatric to adult CHD health care has received increasing attention in recent years. While transfer describes the single event of moving from pediatric to adult services, transition describes the process of preparing individuals for additional health management responsibilities associated with adulthood.[43] For patients, the process of transition entails becoming knowledgeable about their health condition and prognosis, acquiring necessary self-management strategies, and becoming their own primary decision maker.[44–46] Patients and their families should be educated about the importance of lifetime cardiac surveillance with CHD care specialists to optimize health outcomes.[45–47]

Health-care transition occurs at a phase of life often characterized by changes in multiple domains including education, relationships, vocation, and living quarters. Adolescents and young adults navigate identity formation, feeling in between childhood and adulthood and exploring new possibilities.[48] The combination of typical life transitions with the transition of health care may be perceived as overwhelming for some patients and their parents. While many youth with CHD express cautious optimism regarding transition,[49] they also might experience the loss of a valued pediatric health-care team, confusion regarding the adult care system, and uncertainty about their future.[46] Preexisting or emerging psychosocial concerns can present additional challenges to smooth transition.[46] As part of the transition process, it is therefore crucial that the health-care teams address patients' concerns about transition, inquire about psychosocial difficulties, and respond to gaps in patients' understanding of their CHD.[45,46]

In 2011, the American Heart Association published a set of guidelines for best practices in the transition from pediatric to adult CHD care.[45] These include recommendations for key stakeholders in the transition process, namely patients, parents or guardians, primary care providers, and both pediatric and adult cardiology teams. The guidelines highlight the importance of inclusion of children in discussions about their diagnosis and treatment in the years preceding transition so that they might establish a foundation for later self-management. Parents are also an important component of the pre-transition process, and one survey found that only half of the parents consider their children to be ready to transition to adult care.[50] Parents and pediatric health-care providers would ideally work together to encourage and support adolescents and young adults to become greater participants in their health-care management. A structured transition curriculum is one strategy to ensure that patients, as developmentally able, become knowledgeable about their diagnosis and treatment history, the dosing and purpose of all medications, knowing when and how to access health care, and the impact of CHD on lifestyle matters including substance use, family planning, and educational and vocational pursuits. The ultimate goal of a transition program is to enable patients to become their own well-informed health-care advocates.

End-of-Life and Advance Care Planning

With improved medical care and interventions, the average age at death of individuals with CHD has shifted into adulthood.[3,4] Discussions about end-of-life care and advance care planning are thus emerging as priorities for the care of adults with CHD. These discussions, however, have an obvious psychosocial impact. In a survey of 123 adults with CHD, 57% reported moderate to extreme concern about death and dying.[51] Despite the importance of advanced care planning and end-of-life care discussions between patients and health-care providers, such discussions rarely occur. A retrospective study of 48 adults with CHD who died while admitted to hospital found that a minority (10%) had documented end-of-life discussions prior to death.[52] Another study of 200 adult outpatients with CHD found that although 78% voiced a desire for their health-care teams to initiate end-of-life care discussion, only 1% reported actually having discussed end-of-life planning with their medical team.[53] Although health-care providers might be hesitant to broach end-of-life planning with patients before life-threatening complications have occurred, a majority (62%) of patients indicated that they were interested in having end-of-life discussions before these complications occur.[53] Patients further identified having trust in their doctors as the most important facilitator to discussing end-of-life matters and advance care planning.[54] In this same study cohort, only 5% of patients had completed advance directives, although 87% recognized the importance of having an advance directive should they be dying and unable to speak for themselves.[55] Guidelines for the management of adults with CHD also advise that that all patients should be encouraged to complete advance directives.[7] Health-care providers have a responsibility to broach the subject of advanced care planning early with their patients to reduce the psychological distress of patients and their families and ensure that patients' wishes for end-of-life care will be honored.

PSYCHOSOCIAL INTERVENTIONS

Given the unique psychosocial concerns of the growing population of adults with CHD, one would expect that adults with CHD would benefit from tailored psychotherapeutic interventions. Unfortunately, very few adults with CHD and mental health concerns receive mental health treatment.[10,27,56,57] Factors that decrease the likelihood of receiving mental health treatment include old age, physical comorbidities, and the combination of persistent negative affect with social inhibition.[56] An additional potential barrier is the potential difficulty that providers have when recognizing the emotional distress of patients.[26]

Despite discouraging rates of mental health treatment, as a group, adults with CHD appear interested in accessing treatment. One study found that half of all surveyed patients reported a high level of interest in receiving psychological treatment or peer support, with one-third expressing particular interest in stress management and/or coping with heart disease.[57] In a qualitative study, focus group participants described a comprehensive approach to mental health care that includes individual and/or group counseling, opportunities to interact with peers with CHD, and psycho-education to understand common psychological reactions to living with CHD.[9]

To date, there are no published randomized controlled trials evaluating psychotherapeutic interventions for individuals with CHD.[58] It is reasonable to presume that cognitive behavioral therapy (CBT) would benefit this patient population. CBT is a short-term, structured treatment in which therapists and patients work collaboratively to challenge and change the maladaptive thinking patterns and behaviors of patients.[59] Meta-analyses have demonstrated CBT's effectiveness at reducing symptoms of anxiety and depression in the general population,[60,61] as well as in populations with health concerns.[62] Other types of therapy may also hold promise for the adult CHD population. For example, mindfulness-based interventions have shown small to moderate effects on depression, stress, and anxiety in individuals with a chronic medical disease including vascular disease.[63,64] These interventions make use of mindfulness meditation, in which participants are encouraged to develop a nonjudgmental awareness of their experience in the present moment.[65] Interpersonal therapy, with a focus on enhancing the quality of an individual's social and interpersonal functioning, has also been proposed as a recommended treatment option for adults with CHD.[66] There is current research evaluating the efficacy of mental health treatment for adults with CHD which will prove informative in determining the most effective treatment approaches in the clinical setting.

CONCLUSIONS

Adults with CHD constitute a distinct health population with unique psychosocial experiences and concerns. Although research findings have indeed been mixed, many studies have found that rates of psychosocial distress are higher among adults with CHD compared to the general population. Patients themselves have expressed strong interest in psychological treatment. Health-care professionals should be mindful of the possibility of psychosocial distress among patients and refer to a mental health professional when appropriate.

References

1. Warnes CA. The adult with congenital heart disease: born to be bad? *J Am Coll Cardiol.* July 5, 2005;46(1):1–8.
2. Marelli AJ, Ionescu-Ittu R, Mackie AS, Guo L, Dendukuri N, Kaouache M. Lifetime prevalence of congenital heart disease in the general population from 2000 to 2010. *Circulation.* August 26, 2014;130(9):749–756.
3. Verheugt CL, Uiterwaal CS, van der Velde ET, et al. Mortality in adult congenital heart disease. *Eur Heart J.* May 2010;31(10):1220–1229.
4. Khairy P, Ionescu-Ittu R, Mackie AS, Abrahamowicz M, Pilote L, Marelli AJ. Changing mortality in congenital heart disease. *J Am Coll Cardiol.* September 28, 2010;56(14): 1149–1157.
5. Nasr VG, Kussman BD. Advances in the care of adults with congenital heart disease. *Semin Cardiothorac Vasc Anesth.* December 26, 2014;19(3):175–186.
6. Kaemmerer H, Bauer U, Pensl U, et al. Management of emergencies in adults with congenital cardiac disease. *Am J Cardiol.* February 15, 2008;101(4):521–525.
7. Warnes CA, Williams RG, Bashore TM, et al. ACC/AHA 2008 guidelines for the management of adults with congenital heart disease: a report of the American College of Cardiology/American Heart Association Task Force on Practice Guidelines (Writing Committee to Develop Guidelines on the Management of Adults With Congenital Heart Disease). Developed in Collaboration With the American Society of Echocardiography, Heart Rhythm Society, International Society for Adult Congenital Heart Disease, Society for Cardiovascular Angiography and Interventions, and Society of Thoracic Surgeons. *J Am Coll Cardiol.* December 2, 2008;52(23):e1–121.
8. Kovacs AH, Sears SF, Saidi AS. Biopsychosocial experiences of adults with congenital heart disease: review of the literature. *Am Heart J.* August 2005;150(2):193–201.
9. Page MG, Kovacs AH, Irvine J. How do psychosocial challenges associated with living with congenital heart disease translate into treatment interests and preferences? A qualitative approach. *Psychol Health.* 2012;27(11):1260–1270.
10. Horner T, Liberthson R, Jellinek MS. Psychosocial profile of adults with complex congenital heart disease. *Mayo Clin Proc.* January 2000;75(1):31–36.
11. Claessens P, Moons P, de Casterle BD, Cannaerts N, Budts W, Gewillig M. What does it mean to live with a congenital heart disease? A qualitative study on the lived experiences of adult patients. *Eur J Cardiovasc Nurs.* March 2005;4(1):3–10.
12. Callus E, Quadri E, Ricci C, et al. Update on psychological functioning in adults with congenital heart disease: a systematic review. *Expert Rev Cardiovasc Ther.* June 2013;11(6): 785–791.
13. Mondesert B, Khairy P. Implantable cardioverter-defibrillators in congenital heart disease. *Curr Opin Cardiol.* January 2014;29(1):45–52.

14. Burke JL, Hallas CN, Clark-Carter D, White D, Connelly D. The psychosocial impact of the implantable cardioverter defibrillator: a meta-analytic review. *Br J Health Psychol.* May 2003;8(Pt 2):165–178.

15. Dunbar SB, Dougherty CM, Sears SF, et al. Educational and psychological interventions to improve outcomes for recipients of implantable cardioverter defibrillators and their families: a scientific statement from the American Heart Association. *Circulation.* October 23, 2012;126(17):2146–2172.

16. Zomer AC, Vaartjes I, Uiterwaal CS, et al. Circumstances of death in adult congenital heart disease. *Int J Cardiol.* January 26, 2012;154(2):168–172.

17. Rutledge T, Reis VA, Linke SE, Greenberg BH, Mills PJ. Depression in heart failure a meta-analytic review of prevalence, intervention effects, and associations with clinical outcomes. *J Am Coll Cardiol.* October 17, 2006;48(8):1527–1537.

18. Kovacs AH, Moons P. Psychosocial functioning and quality of life in adults with congenital heart disease and heart failure. *Heart Fail Clin.* January 2014;10(1):35–42.

19. Kovacs AH, Silversides C, Saidi A, Sears SF. The role of the psychologist in adult congenital heart disease. *Cardiol Clin.* November 2006;24(4):607–618. vi.

20. Marino BS, Lipkin PH, Newburger JW, et al. Neurodevelopmental outcomes in children with congenital heart disease: evaluation and management: a scientific statement from the American Heart Association. *Circulation.* August 28, 2012;126(9):1143–1172.

21. Kantoch MJ, Eustace J, Collins-Nakai RL, Taylor DA, Bolsvert JA, Lysak PS. The significance of cardiac surgery scars in adult patients with congenital heart disease. *Kardiol Pol.* January 2006;64(1):51–56. discussion 57–58.

22. Gantt LT. Growing up heartsick: the experiences of young women with congenital heart disease. *Health Care Women Int.* July–September 1992;13(3):241–248.

23. Luyckx K, Goossens E, Van Damme C, Moons P. Identity formation in adolescents with congenital cardiac disease: a forgotten issue in the transition to adulthood. *Cardiol Young.* August 2011;21(4):411–420.

24. Oechslin EN, Harrison DA, Connelly MS, Webb GD, Siu SC. Mode of death in adults with congenital heart disease. *Am J Cardiol.* November 15, 2000;86(10):1111–1116.

25. Kourkoveli P, Rammos S, Parissis J, Maillis A, Kremastinos D, Paraskevaidis I. Depressive symptoms in patients with congenital heart disease: incidence and prognostic value of self-rating depression scales. *Congenit Heart Dis.* 2014;10(3):240–247.

26. Bromberg JI, Beasley PJ, D'Angelo EJ, Landzberg M, DeMaso DR. Depression and anxiety in adults with congenital heart disease: a pilot study. *Heart Lung.* March–April 2003;32(2):105–110.

27. Kovacs AH, Saidi AS, Kuhl EA, et al. Depression and anxiety in adult congenital heart disease: predictors and prevalence. *Int J Cardiol.* October 2, 2009;137(2):158–164.

28. Popelova J, Slavik Z, Skovranek J. Are cyanosed adults with congenital cardiac malformations depressed? *Cardiol Young.* July 2001;11(4):379–384.

29. Muller J, Hess J, Hager A. General anxiety of adolescents and adults with congenital heart disease is comparable with that in healthy controls. *Int J Cardiol.* April 30, 2013;165(1):142–145.

30. van Rijen EH, Utens EM, Roos-Hesselink JW, et al. Longitudinal development of psychopathology in an adult congenital heart disease cohort. *Int J Cardiol.* March 18, 2005;99(2):315–323.

31. Utens EM, Bieman HJ, Verhulst FC, Meijboom FJ, Erdman RA, Hess J. Psychopathology in young adults with congenital heart disease. Follow-up results. *Eur Heart J.* April 1998;19(4):647–651.

32. Cox D, Lewis G, Stuart G, Murphy K. A cross-sectional study of the prevalence of psychopathology in adults with congenital heart disease. *J Psychosom Res.* February 2002;52(2):65–68.

33. Freitas IR, Castro M, Sarmento SL, et al. A cohort study on psychosocial adjustment and psychopathology in adolescents and young adults with congenital heart disease. *BMJ Open*. 2013;3(1).
34. Latal B, Helfricht S, Fischer JE, Bauersfeld U, Landolt MA. Psychological adjustment and quality of life in children and adolescents following open-heart surgery for congenital heart disease: a systematic review. *BMC Pediatr*. 2009;9:6.
35. Kantor PF, Andelfinger G, Dancea A, Khairy P. Heart failure in congenital heart disease. *Can J Cardiol*. July 2013;29(7):753–754.
36. Burker EJ, Evon DM, Marroquin Loiselle M, Finkel JB, Mill MR. Coping predicts depression and disability in heart transplant candidates. *J Psychosom Res*. October 2005;59(4):215–222.
37. Enomoto J, Nakazawa J, Mizuno Y, Shirai T, Ogawa J, Niwa K. Psychosocial factors influencing mental health in adults with congenital heart disease. *Circ J*. 2013;77(3): 749–755.
38. Ong L, Nolan RP, Irvine J, Kovacs AH. Parental overprotection and heart-focused anxiety in adults with congenital heart disease. *Int J Behav Med*. September 2011;18(3): 260–267.
39. Rietveld S, Mulder BJ, van Beest I, et al. Negative thoughts in adults with congenital heart disease. *Int J Cardiol*. November 2002;86(1):19–26.
40. Balon YE, Then KL, Rankin JA, Fung T. Looking beyond the biophysical realm to optimize health: results of a survey of psychological well-being in adults with congenital cardiac disease. *Cardiol Young*. October 2008;18(5):494–501.
41. Foster E, Graham Jr TP, Driscoll DJ, et al. Task force 2: special health care needs of adults with congenital heart disease. *J Am Coll Cardiol*. April 2001;37(5):1176–1183.
42. Tong EM, Sparacino PS, Messias DK, Foote D, Chesla CA, Gilliss CL. Growing up with congenital heart disease: the dilemmas of adolescents and young adults [see comment]. *Cardiol Young*. July 1998;8(3):303–309.
43. A consensus statement on health care transitions for young adults with special health care needs. *Pediatrics*. December 2002;110(6 Pt 2):1304–1306.
44. Gurvitz M, Saidi A. Transition in congenital heart disease: it takes a village. *Heart*. July 2014;100(14):1075–1076.
45. Sable C, Foster E, Uzark K, et al. Best practices in managing transition to adulthood for adolescents with congenital heart disease: the transition process and medical and psychosocial issues: a scientific statement from the American Heart Association. *Circulation*. April 5, 2011;123(13):1454–1485.
46. Kovacs AH, McCrindle BW. So hard to say goodbye: transition from paediatric to adult cardiology care. *Nat Rev Cardiol*. January 2014;11(1):51–62.
47. Yeung E, Kay J, Roosevelt GE, Brandon M, Yetman AT. Lapse of care as a predictor for morbidity in adults with congenital heart disease. *Int J Cardiol*. March 28, 2008;125(1):62–65.
48. Arnett JJ. Emerging adulthood: what is it, and what is it good for? *Child Dev Perspect*. 2007;1(2):68–73. 2015-02-24.
49. Moons P, Pinxten S, Dedroog D, et al. Expectations and experiences of adolescents with congenital heart disease on being transferred from pediatric cardiology to an adult congenital heart disease program. *J Adolesc Health*. April 2009;44(4):316–322.
50. Clarizia NA, Chahal N, Manlhiot C, Kilburn J, Redington AN, McCrindle BW. Transition to adult health care for adolescents and young adults with congenital heart disease: perspectives of the patient, parent and health care provider. *Can J Cardiol*. September 2009;25(9):e317–322.
51. Harrison JL, Silversides CK, Oechslin EN, Kovacs AH. Healthcare needs of adults with congenital heart disease: study of the patient perspective. *J Cardiovasc Nurs*. November– December 2011;26(6):497–503.

III. PSYCHOLOGICAL PROFILES AND PROCESSES

52. Tobler D, Greutmann M, Colman JM, Greutmann-Yantiri M, Librach LS, Kovacs AH. End-of-life care in hospitalized adults with complex congenital heart disease: care delayed, care denied. *Palliat Med.* January 2012;26(1):72–79.
53. Tobler D, Greutmann M, Colman JM, Greutmann-Yantiri M, Librach LS, Kovacs AH. End-of-life in adults with congenital heart disease: a call for early communication. *Int J Cardiol.* March 22, 2012;155(3):383–387.
54. Greutmann M, Tobler D, Colman JM, Greutmann-Yantiri M, Librach SL, Kovacs AH. Facilitators of and barriers to advance care planning in adult congenital heart disease. *Congenit Heart Dis.* July–August 2013;8(4):281–288.
55. Tobler D, Greutmann M, Colman JM, Greutmann-Yantiri M, Librach SL, Kovacs AH. Knowledge of and preference for advance care planning by adults with congenital heart disease. *Am J Cardiol.* June 15, 2012;109(12):1797–1800.
56. Schoormans D, Sprangers MA, van Melle JP, et al. Clinical and psychological characteristics predict future healthcare use in adults with congenital heart disease. *Eur J Cardiovasc Nurs.* 2016;15(1):72–81.
57. Saidi A, Kovacs AH. Developing a transition program from pediatric-to adult-focused cardiology care: practical considerations. *Congenit Heart Dis.* July 2009;4(4):204–215.
58. Lane DA, Millane TA, Lip GY. Psychological interventions for depression in adolescent and adult congenital heart disease. *Cochrane Database Syst Rev.* 2013;10:CD004372.
59. Beck JS. *Cognitive Behavior Therapy: Basics and Beyond.* 2nd ed. New York, NY: Guilford Press; 2011.
60. Butler AC, Chapman JE, Forman EM, Beck AT. The empirical status of cognitive-behavioral therapy: a review of meta-analyses. *Clin Psychol Rev.* 2006;26(1):17–31. 2015-02-24.
61. Hofmann SG, Smits JAJ. Cognitive-behavioral therapy for adult anxiety disorders: a meta-analysis of randomized placebo-controlled trials. *J Clin Psychiatry.* 2008;69(4):621–632. 2015-02-24.
62. Osborn RL, Demoncada AC, Feuerstein M. Psychosocial interventions for depression, anxiety, and quality of life in cancer survivors: meta-analyses. *Int J Psychiatry Med.* 2006;36(1):13–34. 2015-02-24.
63. Abbott RA, Whear R, Rodgers LR, et al. Effectiveness of mindfulness-based stress reduction and mindfulness based cognitive therapy in vascular disease: a systematic review and meta-analysis of randomised controlled trials. *J Psychosom Res.* 2014;76(5):341–351. 2015-02-24.
64. Bohlmeijer E, Prenger R, Taal E, Cuijpers P. The effects of mindfulness-based stress reduction therapy on mental health of adults with a chronic medical disease: a meta-analysis. *J Psychosom Res.* 2010;68(6):539–544. 2015-02-24.
65. Kabat-Zinn J. *Full Catastrophe Living: Using the Wisdom of Your Body and Mind to Face Stress, Pain, and Illness.* 15th anniversary ed. New York, NY: Delta Trade Paperback/Bantam Dell; 2005.
66. Morton L. Can interpersonal psychotherapy (IPT) meet the psychosocial cost of life gifted by medical intervention. *Couns Psych Rev.* 2011;26(3):75–86.

PART IV

INTERVENTIONS

The Congenital Heart Disease Intervention Program (CHIP) and Interventions in Infancy

N. Doherty

Western Health and Social Care Trust, Derry, Northern Ireland

C. McCusker

The Queens University of Belfast, Belfast, Northern Ireland; The Royal Belfast Hospital for Sick Children, Belfast, Northern Ireland

OUTLINE

ADJUSTMENT IS A FAMILY AFFAIR

In chapter "A Family Affair" we highlighted how parents and families of children with congenital heart disease (CHD) are at increased risk for emotional difficulties, stress, and family strain. This is important as we have also highlighted the predictive significance of parent and family factors for child neurodevelopmental and especially behavioral outcomes (see chapter: "Is There a Behavioral Phenotype for Children With Congenital Heart Disease?"). Simply put, children with significant CHD fare much better in more cohesive families, with well-adjusted parents who cope with, rather than excessively worry about, their child's condition. Indeed the model outlined in chapter "Is There a Behavioral Phenotype for Children With Congenital Heart Disease?" highlights that such factors are at least, if not more, important as disease and surgical factors in understanding and improving outcomes. To date, however, most efforts have been directed at refining and advancing medical and surgical interventions. While survival rates have indeed improved, psychosocial outcomes have remained stubbornly unchanged.[1] It is time that more concerted efforts were made to strengthen the buffering capacity of the family to minimize the negative impact of the disease.

When we began to look at developing such an intervention program at our center in Belfast in the early 2000s, most evidence-based psychological interventions in pediatrics pertained to disease management strategies such as pain management and medical adherence.[2] Although important, if adjustment is indeed a family affair, interventions need to be targeted at parents and the family to optimize outcomes for the child with CHD. This was the fundamental premise of our *Congenital Heart disease Intervention Program* (CHIP).

THEORETICAL PRINCIPLES

To date, suggested interventions for children with CHD have been largely based on clinical consensus, and generalizing what we know is effective more generally to work with this population. Thus Le Roy and colleagues in 2003 outlined a range of helpful interventions to prepare children for invasive cardiac procedures, underpinned by the general literature on preparation for medical procedures and informed by Piagetian principles of child cognitive development.[3] More recently in 2012, Marino and colleagues issued a consensus statement, endorsed by the *American Heart Association*, recommending regular developmental screening, evaluation, and cognitive remediation given the now established high risk for neurodevelopmental sequelae in this population.[4] However, the CHIP project has been the first systematic, and theoretically grounded, controlled study of psychological interventions for children with CHD and their families.

The importance of a sound theoretical model of how children adjust to any chronic illness, supported by empirical evidence, is crucial for the development of tailored and targeted interventions. The emergent behavioral phenotype, outlined in chapter "Is There a Behavioral Phenotype for Children With Congenital Heart Disease?" suggests, for example, that interventions aimed at promoting personal and interpersonal competencies are more important than mood management interventions per se. Most fundamentally, the theoretical models, and associated evidence, outlined in chapters "Neurodevelopmental Patterns in Congenital Heart Disease Across Childhood: Longitudinal Studies From Europe" and "Is There a Behavioral Phenotype for Children With Congenital Heart Disease?" suggest that intervention targeted at the level of the parent and family will optimize neurodevelopmental and behavioral outcomes for the child.

Our conclusions with respect to children with CHD and their families fit best within family resource and systems models in the broader pediatric literature.[5,6] These models essentially see the physical disease as a stressor to which the child and family need to adapt. Although disease factors (eg, severity, impairment, and visibility) and child factors (eg, age, gender, and cognitive capacity) are important, this adaptation is mediated by what Thompson and colleagues describe as "transactions."[5] These transactions will include those between the health-care system and the child and family (eg, information giving, preparation, and support), but more importantly will occur between the child and family (and especially the mother). Thus the better able the parents are at containing stress, problem solving, promoting independence, and supporting the child through the developmental transitions, challenged by the disease, the better the outcomes for the child. Interventions, therefore, should focus on promoting such attributes in the parent.

The CHIP project was informed by two other principles in pediatric psychology. First, the importance of specific developmental transitions was recognized. Drotar highlights the significance of specific points in the life cycle of the child and family where chronic illness may become particularly disruptive.[7] Successful negotiation of these stages—eg, birth, entering school, and adolescence—is likely to be especially important in building resilience or, conversely, conferring risk. Thus the interventions outlined in this section are targeted at, and tailored to, such key developmental periods. Secondly, while children with chronic health conditions who get referred to psychological services may very well have complex intervention needs, what we are talking about here is secondary prevention—ie, interventions with a population *at risk* for maladjustment, but who may not yet have become so severe that they reach criteria for referral to psychological services. Thus CHIP interventions are about promoting resilience, rather than treating psychopathology, and are relatively brief in scope and nature. This is doubly important as the feasibility of psychological interventions for pediatric populations and their families may have been challenged by economic constraints and the practical challenge of attending sessions for parents whose capacities may be depleted by medical and surgical demands.[7]

AN OVERVIEW OF CHIP

In the early 2000s we began recruiting families into two arms of our study interventions—*CHIP–Infant* and *CHIP–School*. Although the specific elements of each, to be described in this and the next chapter, were tailored to the key developmental challenges for each cohort, common intervention themes underpinned both.

First, we had been impressed with the promise that James Varni had demonstrated for applying problem-solving therapy[8] to children with cancer and their mothers.[9] Although problem-solving therapy has been shown in 2014 to have the strongest evidence base for family interventions in pediatric psychology (including the evidence of the CHIP studies),[10] when we began our work this evidence base was in its infancy. Nevertheless, it appeared consistent with our family-oriented premise in that it trains parents to take an active, solution-focused, approach to managing the worries, fears, and developmental challenges posed by the disease. This not only minimizes distress associated with helplessness and fear, but empowers parents with strategies to optimize parenting and child development. What was particularly attractive about problem-solving therapy was that while it focuses on pertinent current problems, it can give parent an approach that should be generalizable to future challenges and problems, beyond their participation in this program.

Thus both CHIP programs developed an intervention we called *Problem Prevention Therapy*. This targeted the challenges faced by parents of recently diagnosed infants and of 4- to 5-year-old children about to start school. Salient challenges varied. In infancy these related to feeding, attachment, infant development, and coming to terms with the diagnosis and its implications. For families of the children about to start school these related to parent management, child information giving, and promoting independence and activity. Each program was underpinned by specific psycho-educational, coaching, and behavioral interventions pertinent to the above issues and supported by parent manuals and DVDs (see following sections). However, these were contextualized within a problem prevention framework which aimed to promote an active, solution-focused, orientation to stress and distress summarized by the **DO ACT** acronym:

- **D**efine the worry/problem and turn it into a goal;
- **O**ption appraise through brainstorming;
- **A**sses the pros and cons of each option;
- **C**hoose and commit to a solution(s);
- **T**ake action and evaluate outcome and impact.

Secondly, both CHIP programs were informed by the principle of *meaning making*. David De Maso, who had led early research highlighting the importance of maternal perceptions for child outcomes in CHD,[11] also argued for narrative therapy interventions to help families ventilate emotion, construct meaning, and engage in subsequent meaningful coping activity including active collaboration in medical care.[12] Before beginning CHIP we complemented our review of empirical data and the literature, by consulting extensively with parents of older children who had survived CHD. De Maso's recommendations echoed their experience. Parents had told us of the overwhelming roller-coaster of unprocessed emotions (grief, loss of the "normal" child, confusion, and uncertainty) they had faced in their children's infancy when the pace of medical and surgical interventions had left little space for the work of adjustment and relationship building. Related, they had told us about feeling disempowered and unsure of themselves in providing a narrative to their developing children about what "was wrong" with them and in integrating such narratives into a desire to promote resilient, active, and independent children. Helping parents ventilate emotions and construct an empowering narrative about their situation for themselves and their child, thus pervaded both CHIP interventions in different ways.

Finally, CHIP was cognizant that the "transactions" thought to be significant for optimal child and family adjustment also applied to transactions with the community health and education systems which would monitor and "look after" the infant and child when they had left the specialist CHD unit. Thus both programs had outreach elements designed to enhance the knowledge and skills of these community professionals in their transactions with the child and family.

THE CHIP–INFANT PROGRAM

The *CHIP–Infant* program was delivered by a pediatric clinical psychologist and a specialist nurse in pediatric cardiology. Families were recruited from the regional pediatric cardiology center and involved infants, born with CHD, who had required surgical interventions in the first weeks of life. Seventy families agreed to participate (of 73 invited) and included infants with a range of significant cyanotic and acyanotic heart defects, corrected and palliated only. The program was delivered either at home or at hospital, depending on the child's care status at the time of the intervention. Although tailored to the specific condition, surgical status, and the specific concerns and context of the family, there were six (1–2h) sessions with a standard intervention framework. These are summarized in Table 9.1.

TABLE 9.1   Summary of CHIP–Infant Sessions

Session Number	Focus/Content
One	Telling the story—emotional ventilation and construction of meaning[a]
Two	Psycho-education and problem prevention therapy[b]
Three	Facilitated exposure to medical equipment, procedures, and professionals
Four	Maternal responsivity training—feeding and neurodevelopment
Five	Maternal responsivity training—feeding and neurodevelopment
Six	Review session and family functioning[c]

[a] *DVD of experienced parent stories provided.*
[b] *CHIP manual introduced.*
[c] *Community health professional forwarded personalized fact sheets.*

The first of the sessions focused on allowing the parents to tell their story and used the principles of narrative therapy. For many of the parents, this was the first time that they had told the story of the diagnosis of their child's heart defect. In doing so the session enabled emotional ventilation and construction of meaning. It permitted parents to speak about their experience from the time of diagnosis, whether ante- or postnatally, through their hospital stay, interventions, and infant progress to date. The psychologist facilitated emotional processing during these sessions and parents typically reported relief at being able to name feelings of grief, guilt, and anxiety as well as have these normalized in a contained environment.

Session two started with psycho-education via the discussion of a "personalized" fact sheet about the child, their condition, and clinical prognosis, formulated by the pediatric cardiologists. These presented information in a rather different way to the standard defect focused fact sheet. The child's name was used throughout the fact sheet with reference made to their specific clinical course and specific aspects of their condition. Presenting health information in this way has been shown to make it more relevant, understandable, and memorable.[13] Problem-solving therapy was introduced in this session as a way to empower parents with active coping skills to reduce worry and distress and promote optimal child development. Facilitated by the psychologist and nurse, parents generated a current "worry inventory" in relation to their child and indeed the family system. Taking one worry as an example, parents were shown how to reframe worries into goals and then, using the stages summarized by the DO ACT acronym outlined above, work through potential solutions to this.

As an example, one typical worry for parents was often how to explain to the other children what was happening to their sibling and why their parents needed to be away at the hospital so much. This was reframed as a goal (eg, *giving them enough information to help them understand but in a reassuring way to minimize worry*). Parents were encouraged to brainstorm all possible options they could think of without any judgment or caution at this stage. The therapists modeled how free-flowing this should be, even if ideas generated initially appeared wild or nonsensical. This aspect of problem-solving therapy is deemed crucial to break feelings of paralysis and helplessness and start engaging in solution-, rather than problem-, focused thinking. A deliberate wide spectrum of options would be generated (eg, from telling them nothing at all and that they were simply having a little holiday with granny and grandpa to giving them all the facts as they had been given to them). Options were revised, refined, moderated, and sometimes combined. The pros and cons of each were carefully worked out, typically in this example so they could see that withholding information or misleading, while superficially attractive and avoidant of distress, would actually exacerbate their children's fears and worries, as well as undermine future trust and security. The options related to being honest with the other children, while having initial costs with respect to alarm, worry, and fear, typically came to be seen as having the greatest pros in terms of, for example, reducing catastrophic thinking, hearing about reparative treatments and not just a fearsome disease, modeling of a coping stance, involving the other children in aspects of care and ultimately bringing the family closer together. Information and skill needs to implement preferred options were identified (eg, psychological knowledge about how children understand and cope with such news, story book materials to aid age-appropriate understanding, and role playing and how to go about it), with the therapists able to intervene here to help the parents overcome these potential hurdles to implementation. Parents would commit to implementation of the identified option, with the therapist encouraging specificity in terms of timing and context and with the evaluation stage taking place at a future session. In addition, parents were informed that we would be applying the same general approach when looking at other specific worries in future sessions (typically related to feeding, bonding, and infant development).

Session three involved a deeper facilitated exposure to the medical equipment, the ward if relevant and the health professionals involved in treating the infant, than occurred in routine care. The purpose here was to reduce fear, increase understanding and perceived self-efficacy in participating in the nursing and medical care of the infant. In some cases the work was in vivo, where the parents were present on the ward. In other cases it was preparatory for future interventions and, for some, added to the narrative therapy described in session one as it helped process and make sense of the medical elements of their story which had already taken place. Thus, although tailored to the specific care and treatment status of the infant, the same central elements applied to all participants. Among these elements there was a discussion about the diagnosis, current treatment, and medications and what parental understanding was of these, with clarification of any erroneously held information. Reducing anxiety and a heightened emotional state within the medical setting helped free parents to think through their thoughts, worries, beliefs, and questions more clearly. The fact sheet delivered in session two was used to support this.

There is a growing literature on the importance of maternal responsivity training with preterm infants.[14] Several programs across the world have been reported which have common elements of parental support, coaching, and training in becoming sensitive and responsive to infant cues and careful care-taking and stimulation activities to promote neurodevelopment and attachment behaviors.[15-17] Outcomes have been equivocal to date in terms of benefits for infant development and improving mother–infant transactions; more consistent positive effects have been found, however, in terms of reducing negative affect in the mother. Such interventions underpinned sessions four and five in the *CHIP–Infant* program.

These focused on specific elements of development including feeding, neurodevelopment, and the parent–infant relationship. Coaching and psycho-education were used alongside the problem prevention therapy tenet of DO ACT. Environmental and contextual cues (eg, soothing voice, singing, and specific strokes) were used to help the infant distinguish interactions for pleasure or feeding versus medical or invasive transactions. Coaching took place with respect to feeding practices and recognizing and responding to infant cues such as breathlessness, hunger, and satiation, and on ensuring appropriate length of feed cycle with reciprocal interactions in between. Preparation for further feeding stages such as weaning was discussed (Picture 9.1). Information was also given on stimulating the sucking reflex when a child was tube fed, and onward contact with speech and language therapists and dieticians were made as appropriate. Tools and methods to stimulate the sensory, motor, and cognitive development of the infant were introduced (eg, kinesthetic, vestibular, and sensory–motor stimulation). Parents were encouraged to utilize these daily, but importantly, at a time when infant behaviors signaled that they were in a calm and responsive mode, rather than overaroused or distressed.

In session six interventions and parental learning points through the program were reviewed and, within the context of the problem-solving stance, emphasized throughout. Particular attention was given to the family system and looking at ways to build resilience in overall family functioning (Picture 9.2). Parents were also able to discuss how to generalize the framework to other worries and concerns.

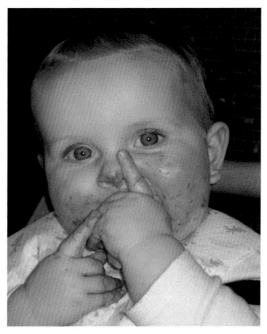

PICTURE 9.1   Photograph of baby spoon-feeding and being all messy.

PICTURE 9.2   Siblings getting involved with care.

IV. INTERVENTIONS

Three additional elements were included in *CHIP–Infant* to support and supplement the direct contact sessions. Importantly, parents were exposed to a DVD after session one which contained parents of older children with CHD, who had been through the stage current participants were at, telling their stories. Feedback suggested these "experienced" parent narratives were quite powerful, not only in normalizing current experiences and facilitating emotional processing, but also in highlighting coping strategies consistent with the content of the program.

Secondly, supporting all the work, and acting as a reference point for the parents, was the CHIP manual which provided summaries of the work of each session as well as additional references and resources. Finally, as noted above, community health professionals, who would be the first point of contact for parents in their communities, were provided with the personalized fact sheets given to parents, with specific information about the child and their condition.

OUTCOMES

The impact of the *CHIP–Infant* program was evaluated within a controlled trial where outcomes across various domains were compared between the intervention group and a control group who had received standard care.[18] Allocation to groups was based on time at admission to the unit so that we did not have cross-contamination of information and experiences across groups. Outcomes were evaluated in relation to both infant development and maternal and family functioning. Although fathers participated in the program, attendance was less reliable and completion of study assessments more patchy than with mothers. Thus information on family-related outcomes came from mothers. Outcomes reported here were comprehensively assessed at 6-month follow-up, with some 7-year follow-up data also attained.

Infant Outcomes

Difficulty with infant feeding is a preponderant concern of parents of children with congenital heart defect. Promoting optimal feeding practice was an area of specific intervention within the CHIP intervention as noted above. Although we did not find any difference between groups in the length of time feeds were taking at 6-month follow-up, there was a significant difference in ease of introducing solids into the diet (16% reported difficulty in intervention group vs. 44% in the control group) and in the rates of breast-feeding at 6-month follow-up.[18] The latter point is particularly important as none of the control group were breast-feeding at 6-month follow-up, whereas 19% of the intervention group continued to incorporate breast-feeding into their babies diet ($X_2 = 5$; $p = 0.03$). These rates are actually above the Northern Ireland average (16%) for the wider population at 6 months after birth.[19] Results from another intervention study,[15] with similar responsivity training in mother and preterm infant dyads, also found a positive secondary impact on rates of breast-feeding.

As noted above, there have been mixed findings regarding the impact of infant intervention programs on neurodevelopment with some finding evidence of positive impact,[20] whereas others[16] have not. In the *CHIP–Infant* program infant development was assessed using the *Bayleys Scales of Infant Development-II* (BSID-II).[21] This is the most widely used scale to measure infant development directly. There were two main subscales: the *Psychomotor Development Index* (PDI) related to aspects of gross and fine motor coordination and the *Mental Development Index* (MDI) which measures the basic cognitive building blocks of higher-order mental abilities such as memory, habituation, vocalizations, language, and problems solving. Each subscale has a normative mean of 100 and a standard deviation of 15.

Table 9.2 summarizes the 6-month follow-up outcomes for the program intervention infants and the control group. We have included data for the total sample and following exclusion of the 11 infants, across both groups, with an additional Down syndrome diagnosis. The relative pattern of findings was consistent. There was no discernible difference between groups on the PDI scale with both groups falling 1–2 standard deviations below the normative mean. This suggests significant developmental delay within this domain consistent with the profiles outlined in chapters "A Longitudinal Study From Infancy to Adolescence of the Neurodevelopmental Phenotype Associated With d-Transposition of the Great Arteries," "Neurodevelopmental Patterns in Congenital Heart Disease Across Childhood: Longitudinal Studies From Europe," and "An Emergent Phenotype: A Critical Review of Neurodevelopmental Outcomes for Complex Congenital Heart Disease Survivors During

TABLE 9.2 Mean PDI and MDI Scores, and Statistical Comparisons Between Groups, at 6-Month Follow-Up

Index Scale	Mean (SD)		F (p)	Partial Eta Squared
	Intervention	Control		
PDI	69.3 (17.1)	71.7 (18.1)	0.24 (0.63)	0.005
MDI	98.4 (11.1)	86.7 (17.7)	5.38 (0.02)	0.101[a]
Syndrome Excluded				
PDI	72.9 (16.5)	75.8 (16.9)	0.72 (0.40)	0.008
MDI	100.3 (10.2)	92.9 (12.1)	4.23 (0.04)	0.098[a]

[a] *Moderate to large effect size.*

Infancy, Childhood, and Adolescence." However, there were statistically significant differences between groups on the MDI which reflects mental development and relies less on motor coordination abilities. These differences, which had clinically significant effect sizes, were maintained even when the children with developmental syndromes were excluded from the analyses. It would appear that the *CHIP–Infant* program conferred a positive benefit on nonmotor aspects of cognitive development in these CHD infants at least at 6-month follow-up.

Family Outcomes

The *CHIP–Infant* program wanted to examine adjustment, coping, and mental health difficulties in the parents, not just because of the direct impact having a child with a chronic condition can have on parents themselves, but also due to the secondary impact via interaction between parent and child (see chapters: "Is There a Behavioral Phenotype for Children With Congenital Heart Disease?" and "A Family Affair"). Mental health in this cohort at program baseline is discussed in chapter "A Family Affair" and in detail in the 2009 article by Doherty et al.[22] We wished to examine the outcomes relating to anxiety, worry, and coping, any changes over time, and in comparison to the control group.

Outcomes regarding maternal anxiety, worry, and coping were examined, using scales widely used in the clinical and research literature—*Spielberger State Anxiety Scale* (form Y),[23] the *Maternal Worry Scale*,[24] and the situation-specific *COPE* (which measures various adaptive- and maladaptive-coping strategies in relation to the stresses of managing CHD in this case).[25] Figs. 9.1 and 9.2 summarize our findings that mothers in the intervention group were manifesting significantly reduced levels of both anxiety and worry, at 6-month follow-up, and in comparison to the control group who had not received the *CHIP–Infant* interventions. Specifically, the percentage of mothers with anxiety levels more than 1 standard deviation from the normative mean (at-risk range) fell from 26% to 3% in the intervention group but only from 30% to 22% in the control group (see Fig. 9.1). Maternal worry levels were comparable at baseline, but at follow-up there was a significant difference with mothers in the intervention group demonstrating a significantly greater reduction in worry (see Fig. 9.2).

Coping and resilience have been increasingly found to be important in studies examining family factors in chronic illness.[26,27] In this study, those mothers who participated in the *CHIP–Infant* program showed a trend on all subscales to have more adaptive coping skills (eg, active vs. passive coping, planning and problem solving, and engagement vs. disengagement) with statistically significant advantages over the control group evident on some (eg, positive reinterpretation and growth).[18]

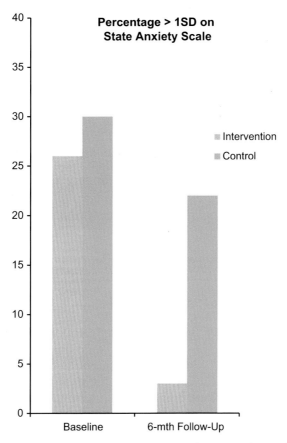

FIGURE 9.1   Percentage of mothers in both groups who were in the at-risk range on the Spielberger state anxiety scale[23] at baseline and 6-month follow-up.

CONCLUSIONS

In *CHIP–Infant* we have gone beyond clarifying the factors which mediate outcomes for children with CHD and their families, and applied this knowledge to a new intervention program aimed at bolstering resilience and reducing risk. We have demonstrated that such interventions can indeed have a positive impact on (A) maternal mental health and coping (which we know impacts on longer-term child outcomes) and (B) infant cognitive development and feeding interactions.

The importance of early intervention in infancy, with at-risk populations, has been increasingly recognized by both public health and government bodies.[28,29] Although the focus has often been at the macro-level with respect to socioeconomics, this chapter has highlighted the positive impact of intervening at the level of the family and parent–infant

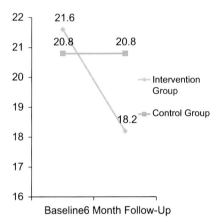

FIGURE 9.2   Mean scores on the maternal worry scale[24] across both groups at baseline and at 6-month follow-up.

interactions. Bolstering resilience and protective factors are likely to have long-term personal, social, and economic advantages in terms of reducing disability and promoting better educational and ultimately occupational inclusion. For our CHD population, this is especially pertinent when one considers the point highlighted by Jackson et al. in their comprehensive review,[30] with regard to the protective bidirectional role of attachment for this population, with more positive health gains evident for securely attached infants with CHD.[31]

The empirical question of whether gains accrued from intervention programs, such as *CHIP–Infant*, can be maintained in the longer term is important. However, few intervention trials in the pediatric psychology literature have follow-up periods even as long as 6 months.

When we looked at our sample more recently at 7 years of age, selective attrition rates compromised our capacity to make meaningful comparisons between our original intervention and control group on many variables.[32] Trends were observed, but statistical power was low and few clear differences were noted. We were able to conclude, however, that those factors we had been able to positively influence when these children were infants (maternal mental health, worry, coping skills, and infant cognitive functioning), predicted outcomes 7 years later for the sample as a whole (intervention and control group). This provides some evidence of longer-term benefit for the intervention albeit by inference. More directly, some evidence was found for statistically significant differences between the original intervention and control group in terms of behavioral adjustment in the children. The small numbers involved in this follow-up study mean that this finding should be interpreted with caution, but results are none-theless promising.

In the next chapter we describe, and review the impact of, the *CHIP–School* program. Incorporating many of the same essential tenets of *CHIP–Infant*, but adapted to the developmental challenges and contexts of early childhood, the importance and benefits of family-oriented interventions in CHD is again our focus.

References

1. Spijkerboer AW, Utens EM, Bogers AJ, Helbing WA, Verhulst FC. A historical comparison of long-term behavioral and emotional outcomes in children and adolescents after invasive treatment for congenital heart disease. *J Pediatr Surg*. 2008;43:534–539.
2. Barlow JH, Ellard DR. The psychosocial well-being of children with chronic disease, their parents and siblings: an overview of the research evidence base. *Child Care Health Dev*. 2006;32:19–31.
3. Le Roy S, Elixson EM, O'Brien P, et al. Recommendations for preparing children and adolescents for invasive cardiac procedures: a statement from the American Heart Association Pediatric Nursing Subcommittee of the Council on Cardiovascular Nursing in collaboration with the Council on Cardiovascular Diseases of the Young. *Circulation*. 2003;108:2550–2564.
4. Marino BS, Lipkin PH, Newburger JW, et al. Neurodevelopmental outcomes in children with congenital heart disease: evaluation and management: a scientific statement from the American Heart Association. *Circulation*. August 28, 2012;126(9):1143–1172.
5. Thompson RJ, Gustafson KE, Hamlett KW, Spock A. Stress, coping and family functioning in the psychological adjustment of mothers of children with cystic fibrosis. *J Pediatr Psychol*. 1992;17:573–585.
6. Robin A, Foster A. *Negotiating Parent-Adolescent Conflict: A Behavioral Family Systems Approach*. New York: Guildford Press; 1998.
7. Drotar D. *Psychological Interventions in Childhood Chronic Illness*. Washington, DC: American Psychology Association; 2006.
8. D'Zurilla TJ, Nezu AM. *Problem-Solving Therapy: A Social Competence Approach to Clinical Intervention*. 2nd ed. New York, NY: Springer; 1999.
9. Varni JW, Sahler OJ, Katz ER, et al. Maternal problem-solving therapy in pediatric cancer. *J Psychosoc Oncol*. 1999;16:41–71.
10. Law EF, Fisher E, Fales J, Noel M, Eccleston C. Systematic review and meta analysis of parent and family-based interventions for children and adolescents with chronic medical conditions. *J Pediatr Psychol*. 2014;39:866–886.
11. De Maso DR, Campis LK, Wypij D, Bertram S, Lipshita M, Freed M. The impact of maternal perceptions and medical severity on the adjustment of children with congenital heart disease. *J Pediatr Psychol*. 1991;16:137–149.
12. De Maso DR, Gonzalez-Heydrich J, Erickson JD, Grimes VP, Strohecker C. The experience journal: a computer-based intervention for families facing congenital heart disease. *J Am Acad Child Adolesc Psychiatry*. 2000;39:727–734.
13. Kreuter M, Holt C. How do people process health information? Applications in an age of individualized communication. *Curr Dir Psychol Sci*. 2001;10:206–209.
14. Benzies KM, Magill-Evans JE, Hayden KA, Ballantyne M. Key components of early intervention programs for preterm infants and their parents: a systematic review and meta-analysis. *BMC Pregnancy Childbirth*. 2013;13:S10–S25.
15. Glazebrook C, Marlow N, Israel C, et al. Randomised trial of a parenting intervention during neonatal intensive care. *Arch Dis Child Fetal Neonatol Ed*. 2007;92:F438–F443.

16. Ravn IH, Smith L, Smeby NA, et al. Effects of early mother-infant intervention on outcomes in mothers and moderately and late preterm infants at age 1 year: a randomized controlled trial. *Inf Behav Dev*. 2012;35:36–47.

17. Meijssen DE, Wold M, Van Bakel H, Koldewijn K, Van Baar A. Maternal attachment representation after very preterm birth and the effects of early intervention. *Inf Behav Dev*. 2011;34:72–80.

18. McCusker CG, Doherty NN, Molloy B, et al. A controlled trial of early interventions to promote maternal adjustment and development in infants with severe congenital heart disease. *Child Care Health Dev*. 2009;36(1):110–117.

19. *Breastfeeding – A Great Start: A Strategy for Northern Ireland 2013–2023*; 2013.

20. Orton J, Spittle A, Doyle L, Anderson P, Boyd R. Do early intervention programmes improve cognitive and motor outcomes for preterm infants after discharge? A systematic review. *Dev Med Child Neurol*. 2009;51:851–859.

21. Bayley N. *Bayley Scales of Infant Development*. 2nd ed. San Antonio, TX: The Psychological Corporation, Harcourt Brace & Company; 1993.

22. Doherty NN, McCusker CG, Molloy B, et al. Predictors of psychological functioning in mothers and fathers of infants born with severe congenital heart disease. *J Reprod Infant Psychol*. 2009;27:390–400.

23. Spielberger CD, Gorsuch RL, Lushene R, Vagg P, Jacobs GA. *Manual for the State-Trait Anxiety Inventory*. California: Consulting Psychologists Press, Inc.; 1983.

24. DeVet KA, Ireys HT. Psychometric properties of the maternal worry scale for children with chronic illness. *J Pediatr Psychol*. 1998;23:257–266.

25. Carver CS, Scheier MF, Weintraub JK. Assessing coping strategies: a theoretically based approach. *J Personal Soc Psychol*. 1989;56:267–283.

26. Kazak A. Pediatric Psychosocial Preventative Health Model (PPPHM): research, practice, and collaboration in pediatric family systems medicine. *Fam Syst Health*. 2006;24(4):381–395.

27. Dale M, Solberg O, Holmstrom H, Ladolt M, Eskedal L, Vollrath M. Mothers of infants with congenital heart defects: wellbeing from pregnancy through the child's first six months. *Qual Life Res*. 2012;21(1):115–122.

28. Public Health Agency, Northern Ireland. Making the Best Start in Life for Newborns and Infants. A 3 Year Framework for Infant Mental Health 2014–2017; 2014.

29. Leadsom A, Field F, Burstow B, Lucas C. The 1001 Critical Days. The Importance of the Conception to Age Two Period. A Cross Party Manifesto.

30. Jackson AC, Frydenberg E, Liang RP-T, Higgins RO, Mirphy BM. Familial impact and coping with child heart disease: a systematic review. *Paediatr Cardiol*. 2015;36(4):695–712.

31. Goldberg S, Simmons RJ, Newman J, Campbell K, Fowler RS. Congenital heart disease, parental stress, and infant-mother relationships. *J Pediatr*. 1991;119(4):661–666.

32. McCusker CG, Armstrong M, Mullen M, Doherty NN, Casey FA. A sibling controlled, prospective, study of outcomes at home and school in children with severe congenital heart disease. *Cardiol Young*. 2013;23:507–516.

Growing Up: Interventions in Childhood and CHIP–School

C. McCusker

The Queens University of Belfast, Belfast, Northern Ireland; The Royal Belfast Hospital for Sick Children, Belfast, Northern Ireland

OVERVIEW

As outlined in chapter "The Congenital Heart Disease Intervention Program (CHIP) and Interventions in Infancy", the fundamental premise of the *Congenital Heart Disease Intervention Program* (CHIP) was that interventions aimed at the parents and families of children with congenital heart disease (CHD) would not only have benefits for their mental health and adjustment, but also for the adjustment and psychosocial functioning of the child with CHD. This premise is supported by empirical evidence presented throughout this book that parental and family functioning is often more important in mediating outcomes for the child than disease or

Congenital Heart Disease and Neurodevelopment
http://dx.doi.org/10.1016/B978-0-12-801640-4.00010-X

surgical factors per se. *CHIP–Infant*, which primarily included problem prevention therapy, construction of meaning, responsivity training, and neurodevelopmental stimulation, was shown in a controlled trial to confer positive advantages for infant cognitive development and maternal mental health. Given the predictive significance of maternal adjustment during infancy for child outcomes years later,[1] this is an important finding.

Nevertheless, as noted in chapter "The Congenital Heart Disease Intervention Program (CHIP) and Interventions in Infancy", the question of whether such benefits are maintained in the longer term is equivocal. It is likely that "booster" sessions will be required and adapted to subsequent developmental stages. *CHIP–School*, although carried out with a different cohort of children with CHD and their families, represents such an adaptation. Drotar[2] suggests timing interventions in pediatric psychology research to key developmental transitions. We designed *CHIP–School* with this principle in mind and focused on the key developmental transition of starting school.

When we spoke with our previous service users (parents of grown-up children with CHD) at the conception and design of CHIP, they echoed what we had seen in the literature about the significance of this developmental transition for children with CHD and their families.[3] Parents spoke about how starting school reactivated all sorts of anxieties about their child's health and capacity to stay safe when removed for large parts of the day from their protection and care. Tensions centered around treating the child "normally" versus anxieties about safe activity levels, tantrums putting a "strain" on the heart and how to talk with the child about their condition and frequent hospital visits. Thus, as *CHIP–Infant* focused on the specific challenges of feeding, attachment, and neurodevelopment, *CHIP–School* adapted problem prevention therapy to incorporate the specific challenges parents of children with CHD starting school faced—activity levels, adapting to school, parenting in the context of CHD, and meaning making for the child. As with *CHIP–Infant*, our fundamental assumption was that if we could address and reduce the barriers to successful adaptation in the parents, we would effectively be promoting healthier adaptation in the child.

CHIP–SCHOOL: A RANDOMIZED CONTROLLED TRIAL[4]

Randomized controlled trials (RCTs) have been recognized as the "gold standard" methodology when evaluating the efficacy of an intervention. The increase in studies using RCT designs was highlighted as crucial in advancing the evidence base for psychological interventions in pediatrics during the 2010s—even to the point where they may be considered "first-line treatments."[5] In psychological research this typically involves random assignment of participants to study versus control groups to (1) assess an intervention against a comparable group who follow the same developmental and clinical trajectory but without the intervention and (2) minimize any potential bias in who gets allocated to what condition. In addition, in RCTs there is further attention to internal validity by ensuring that treatments are delivered in a standardized way. *CHIP–School* was such an RCT.

Recruitment operated across a 3-year period at the regional center for pediatric cardiology at the *Royal Belfast Hospital for Sick Children* in Northern Ireland. We invited all 4- to 5-year-old children and their families who had undergone an invasive procedure to correct or palliate a major heart condition in infancy. Thus we had children who had a range of cyanotic and acyanotic conditions and who had open and closed/transcatheter interventions (see chapter: Congenital Heart Disease: The Evolution of Diagnosis, Treatments, and Outcomes). As noted in chapter "Is There a Behavioral Phenotype for Children with Congenital Heart Disease?," the behavioral phenotype of those with comorbid developmental syndromes was likely to vary, and thus these children were not invited to participate in this particular trial (unlike *CHIP– Infant* where such differences were not deemed significant to the targeted interventions at that developmental stage). We had 90 of 149 accepting the invitation to participate with those who opted in being no different to those who did not in terms of age, gender, socioeconomic status, and disease and surgical features. Following baseline assessments conducted in the 3 months prior to starting school,[6,7] our families were randomly assigned to either the intervention or the control group and follow-up assessments were conducted for both groups at the end of the child's 1st year at school (about 10 months later). The control group received treatment as usual over the time period of the study. Twenty-two families were lost to follow-up (withdrew or failed to appoint within the time frame), and we were left with 33 in the intervention and 35 in the control group. Although those retained across the year were similar to those lost to follow-up on demographic, disease, and surgical features, it was noted that those lost to follow-up had higher levels of child behavior difficulties and maternal worry at baseline. The relevance and significance of this are considered in the following sections.

CHIP–SCHOOL: INTERVENTIONS

CHIP–School was similar to *CHIP–Infant* in its theoretical basis, its emphasis on problem prevention therapy, meaning making, outreach to community services, and its relatively brief duration of about 8 h.

However, it was structured differently (primarily involving group, in addition to individual, interventions) and wove psycho-education and behavioral interventions into problem prevention therapy which were specific to this particular developmental stage for the child and family. Again sessions were delivered by clinical psychologists, pediatric cardiologists, and a specialist pediatric cardiology nurse. In terms of structure there were four components.

First, 9–12 parents per group attended a one-day 5-hour workshop held in the 1st month of the child's 1st year at school. A group format was deemed economical but, more importantly, most appropriate here to the nature of the interventions especially the process of meaning making with other parents facing similar challenges. Following an outline of the day and rationale for psychological interventions, the workshop had three sections:

- *Problem Prevention Therapy*: Through small group work, followed by larger group discussion, parents were enabled to tell their stories to each other with the facilitator helping parents to make sense of their shared experience as part of the "normal journey" of parenting a child with significant CHD. This appeared, and was later noted by parents (see following paragraphs), to be an extremely therapeutic and engaging process in itself. However, a key point of this process was to then generate a collection of salient, often previously unarticulated, worries and fears about their child with CHD, themselves as parents and their families. These were collated and clustered as part of the meaning making process and typically related to managing challenging behaviors in a child with a "weak" heart, promoting independence, safe activity levels, how to talk with the child about their condition, distinguishing functional from "real" sick behavior, and ensuring that the other siblings were not neglected within the family. Problem prevention therapy was introduced as outlined in chapter "The Congenital Heart Disease Intervention Program (CHIP) and Interventions in Infancy," and the workshop facilitators modeled working through a specific worry by turning it into a goal and using the subsequent problem-solving stages summarized by the DO ACT acronym, as outlined in chapter "The Congenital Heart Disease Intervention Program (CHIP) and Interventions in Infancy." Parents then had a chance to work together in small groups to apply the problem prevention stages to a problem with which they could all identify. It was noted that later sessions during the workshop, and their one-to-one sessions with the team, would provide further knowledge and skills for them to weave into future problem solving.
- *Psycho-education*: In this section a review of diagnoses, treatments, prognoses, and future courses were discussed. The emphasis was on normalization, and promoting activity and resilience in the

child. Future issues which had also arisen in the worry generation session were discussed with options considered within the problem prevention framework (eg, with respect to health insurance, fertility, and pregnancy). To supplement this group session, all participants were provided with a personalized fact sheet about their child and their condition, formulated by the pediatric cardiologists. This was different from the general, condition-related, fact sheets available in the unit to the control group. The emphasis in this health information was on the child—rather than on the condition—and the guidance related to the child's specific manifestation of the condition and as informed by treatment, progress, and associated information gleaned from all these to date. Such individualized fact sheets were used to render information more personally relevant, credible, and memorable.[8]

- *Parenting the Child with CHD*: This final session of the workshop focused on parenting skills, generally and specifically in relation to the child with CHD, with links again being made to the worry list generated earlier in the day. Topics covered were tailored to the specific issues raised by each workshop group, but typically included effective communication and reinforcement strategies, preparation for medical procedures, information giving, and looking after the rest of the family. A self-help manual was provided to cover topics in more detail and to help maintain and generalize gains. Across scenarios facilitators highlighted unhelpful beliefs such as *my child has been through too much to worry about discipline… my child is not fit to participate in physical education… I need to protect my child from the reality of her condition…* and used cognitive therapy principles to gently question and challenge.

The remaining components of *CHIP–School* involved one-to-one sessions with each family. The second component involved a behavioral experiment related to the worry about safe activity levels and was informed by our studies, and those of others, highlighting how maternal worry and distorted perceptions are associated with reduced activity levels in children with CHD.[7,9] At these sessions parents and their child attended a specially designed bicycle exercise stress test overseen by the pediatric cardiologist and the nurse. This was framed as an objective review and check on their child's exercise capacity. However, the primary aim was to actually show parents, under safe and controlled conditions, how much vigorous exercise capacity the child was capable of without any adverse impact on the heart and thus challenge unhelpful assumptions and beliefs. The child engaged in bicycling with equipment lights coming on progressively to reflect increased effort and duration of exercise. Continuous ECG monitoring was conducted during the exercise and the recovery period with cardiologists drawing attention to the continuing normality of ECG rhythms, despite the increased heart rate. Observations and conclusions were discussed with families at the end of the test and linked to decision making about activity levels within the problem prevention framework.

The third and final family contact session occurred several weeks after the bicycle exercise test. This involved a session with one of the project psychologists, aimed at reviewing and consolidating the learning accrued during the workshop and the bicycle test. The application of problem prevention therapy to their own previously identified worries was reviewed together with psycho-education promoting the psychological health of siblings and parents of children with a chronic illness.

Finally and in common with *CHIP–Infant* the current program involved outreach to community health and education services. The personalized fact sheets were circulated to general practitioners, community pediatricians, and schoolteachers—the three key groups that our service user consultants during the project development had suggested had often been a source of confusion and unease for them due to their own relative uncertainty and anxiety about managing children with complex CHD. The personalized fact sheets aimed to be helpful in affirmation of a normalization ethos in these first-line community professionals who would interface regularly with the child and family. A summary of all *CHIP–School* interventions is outlined in Fig. 10.1.

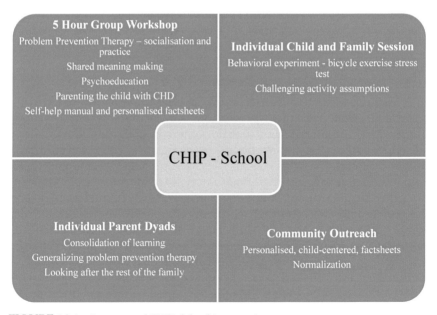

FIGURE 10.1   Summary of CHIP–School interventions.

INTERVENTION OUTCOMES AND COMMENTS

Family-based interventions for children with chronic illnesses and their families have tended to show relatively modest outcomes to date.[10–12] In this context the outcomes for *CHIP–School* appear very promising.[4]

Looking first at maternal- and family-reported outcomes (returns at follow-up were insufficient to analyze paternal changes and patterns). Although worry, as measured by the *Maternal Worry Scale*[13] remained stable in both groups across time, the data depicted in Fig. 10.2 highlight a significant interaction effect between groups and across time for maternal mental health. This was assessed using the *Brief Symptom Inventory*,[14] and highlights that while standardized T scores dropped in the intervention group across the child's 1st year at school, levels of psychopathology actually increased in the control group during this key developmental transition for the child. The interaction was both statistically significant and represented a large clinical effect size.[15] This was consistent with maternal reports about their health since project baseline. Both groups reported comparable levels of physical illness in the 10-month period (about 12% reported experiencing physical health complaints); while only 3% of the intervention group reported having experienced mental health difficulties, this figure was 21% in the control group.

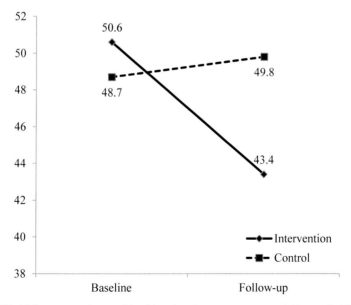

FIGURE 10.2 Maternal mental health at baseline assessment and 10-month follow-up, mean GSI T scores. *GSI*, general severity index—summary scores from the brief symptom inventory.

The *Impact on Family Scale* is a self-report measure of the perceived impact of having a child with a chronic illness on family functioning across four domains: *financial*; *family strain* (restrictions imposed on the family as a whole); *personal strain* (burden of care); and *mastery* (illness experiences bringing the family closer).[16] As summarized in Table 10.1 there was evidence for a positive impact of the intervention on both family strain and personal strain (as rated by mothers) with levels of both falling in the intervention group but rising in the control group across the 10-month period of the study. These interaction effects were significant.

These findings are very encouraging. Results derived from this RCT suggest that participation in *CHIP–School* had positive benefits for maternal mental health and family functioning—the key factors which have been reliably shown to be associated with long-term outcomes for children with CHD. The question raised, however, was would similar benefits be shown for child outcomes within the time frame of the study as they had in *CHIP–Infant*.

Parent and family interventions for children with chronic illness have tended to find a stronger impact on parental and family functioning than for child outcomes per se,[11,12] and to some extent this was also our experience. Disappointingly, no differences were found between groups in terms of teacher ratings of behavioral adjustment.[4] We did find rates of clinically significant behavior problems as measured by the *Child Behavior Checklist*[17] to be almost halved across time in the intervention group (from 21% at project baseline to 12% at 10 month follow-up), whereas rates actually increased slightly in the control group (from 11% to 14%). However, this effect was not significant and may have been influenced by two factors. First, the greater attrition rate involving families of children with higher levels of behavior problems undoubtedly reduced the statistical variability

TABLE 10.1   Impact on Family—Changes Across Time in the Intervention and Control Groups

Impact on Family — Mean Scores (SD)	Intervention Group (N=33)		Control Group (N=35)		F (df)	p (Partial Eta Squared)
	T1	T2	T1	T2		
Family strain	17.2 (4.1)	15.8 (4.4)	16.1 (4.3)	17.0 (3.7)	5.44[a] (1, 61)	0.02 (0.08)
Financial impact	7.8 (2.5)	8.0 (3.5)	7.3 (2.5)	8.0 (2.5)	0.59[a] (1, 61)	0.44 (0.01)
Personal strain	12.0 (3.1)	11.3 (3.3)	10.9 (3.7)	12.1 (3.5)	7.18[a] (1, 61)	0.01 (0.11)
Mastery	15.9 (1.9)	16.2 (1.9)	16.1 (2.4)	16.3 (2.0)	0.13[a] (1, 61)	0.72 (0.002)

[a]*F statistic for the interaction effect.*

to detect change and indeed power. Secondly, however, it is possible that a longer time period is required for the positive impact the program had on parents and families to translate into positive outcomes for the child with CHD. Intervention which is related to prevention of difficulties is likely to have a longer latency for impact than interventions which are treating currently active and significant levels of maladjustment. Nevertheless, the drop in clinically significant levels of behavior problems in the intervention group is promising, and further follow-up of this sample will help clarify if a differential impact in comparison to the control group does become more apparent with time.

Clearer benefits were, however, evident for two important outcomes which had been directly targeted during the problem prevention therapy interventions. Specifically, children in the intervention group missed fewer days from school during their 1st year, based on data from school records. Days missed were on average twice as high in the control group compared to the intervention group (10 vs. 5) as outlined in Table 10.2. In addition parents were asked to record the number of days they perceived their child to be "sick" and in need of unscheduled medical attention in the 3 months prior to project baseline assessment and at the follow-up period at the end of the 1st year at school. An interaction effect was evident here with the average number of days dropping from 4 to 3 in the intervention group but doubling across the same time period in the control group from 4 to 8 (see Table 10.2).

Together these findings are encouraging with a number of positive outcome indicators for the mother, the family, and the child. Finally, however,

TABLE 10.2 Child Outcomes—Days Sick and Days Off School

Outcome Measure	Intervention Group (N=33)		Control Group (N=35)		F/t (df)	p (Partial Eta Squared)
	T1	T2	T1	T2		
Days Sick						
Mean days sick past 3 months (SD)	4.4 (6.4)	3.3 (3.6)	3.7 (7.3)	7.7 (10.6)	5.61[a] (1, 65)	0.02 (0.08)
School Functioning						
Mean days off school since start (SD)	–	4.9 (4.6)	–	9.7 (11.1)	2.30[b] (65)	0.02 (0.06)

[a]F statistic for the interaction effect.
[b]t statistic between groups.

we were interested in the subjective experience of the families who participated in *CHIP–School*. Formal outcomes as reported above are important for assessing program *efficacy*, but the *effectiveness* of any intervention must also be evaluated with respect to acceptability and feasibility of the interventions offered. Parents completed an anonymous open-ended and structured questionnaire 2–3 weeks after their participation in the intervention program. There was a 90% return rate which in itself suggested high levels of engagement with the program. Open-ended responses were subjected to thematic analysis[4] with the most common themes being related to improved knowledge and parenting skills, appreciation of the shared stories and meaning making in the group workshop, attitude change regarding the child's exercise capacity, and appreciation of the CHIP manual, personalized fact sheets, and information to schools. These subjectively perceived benefits were consistent with the theoretical assumptions about *how* we would affect change on the formal outcome measures reported above and are thus important findings. Moreover, parental reports here shed some light on the absence of change on the formal maternal worry scale noted in the preceding paragraphs. Several parents explicitly noted that they would continue to "worry," despite feeling "better equipped" now to deal with this worry. The structured questions pertaining to specific program elements suggested high levels of acceptability and positive experiences across all components.[4]

CONCLUSIONS AND CONTEXT

There is a gap between recognizing the importance of psychological factors in mediating outcomes for children with CHD and their families and demonstrating the efficacy and effectiveness of interventions which putatively follow. Indeed while clinical consensus and guidelines recommend access to psychological interventions for children with CHD and other conditions,[18–21] the evidence base for what works for whom, and at what point, in pediatric psychology is generally in its infancy.[12] The CHIP project has been the first major trial to translate evidence from the knowledge base about phenotypes and etiologies involved in CHD into efficacious and effective interventions.

Psycho-education, which shares findings about risk and protective factors might be thought of as a first-stage intervention in this domain, and undoubtedly this can be helpful for many families—especially those who have the capacity to independently translate and apply this to personal circumstances.[22] However, this alone has not proven effective in formal evaluation of secondary prevention programs for families of children with other chronic illnesses.[10] Rather, it appears that more experiential and structured therapeutic protocols are required to affect clinically significant change. *CHIP–School* has woven a comprehensive range of therapeutic components into a relatively brief intervention program. This has included problem-solving therapy, meaning making, parent skills training targeted at developmental transitions challenged by the disease, cognitive behavioral interventions to shift assumptions about activity capacity, providing information which is child (rather than condition) specific, and outreach to community services.

Importantly, *CHIP–School* has overcome many of the limitations currently identified in the literature for pediatric psychology interventions.[12,23] The formulation of the program has been theoretically grounded in a transactional model which reflects etiological pathways observed in the literature. The focus has been on promoting competencies and skills rather than on mood regulation per se, given the relative significance of the former as the behavioral phenotype (see chapter: Is There a Behavioral Phenotype for Children with Congenital Heart Disease?). A control group has been included, and outcomes have been evaluated over a 10-month period and under RCT conditions. Thus our findings appear robust.

As has been generally observed in such interventions with other pediatric populations,[12] the clearest benefits have been for the parents (and in particular the mothers) in terms of reductions in distress, personal and family strain, and in better mental health. However, we have also demonstrated a trend for improved behavioral adjustment in the child, and significantly reduced child's "sick" behaviors (as reflected in unscheduled health-care utilization) and fewer days missed at school. Finally, we have demonstrated that not only is such a program acceptable to, and appreciated by, parents at this stage in the child's development and illness trajectory, but the themes identified from their feedback about their experiences were consistent with the theoretical mechanisms of action proposed at study design.

CHIP–School and *CHIP–Infant* present a new intervention model for parents of children with CHD which is generalizable across different developmental stages and transitions. Central to both programs is problem-solving therapy for parents. The fact that this therapy has also shown promise in controlled intervention trials with parents of children with cancer,[24,25] diabetes,[11] and traumatic brain injury[22] suggests that this is likely to be an important element of the program. However, the CHIP program incorporates other, theoretically based and empirically driven, intervention elements including meaning making and narrative therapies and outreach to community services. Moreover, it weaves within the problem-solving framework interventions which we suggest are important to help families successfully negotiate specific challenges of a given developmental transition—eg, responsivity training and neurodevelopmental stimulation for parents of the at-risk infants, and parenting skills and activity promotion for parents of those children about to enter school.

As the field of pediatrics evolves beyond medical interventions to increase survival and minimize impairment and toward more multidisciplinary approaches to promote better quality of life and reduced disability, programs such as CHIP will become increasingly important. Interest in affecting positive change for the child through indirect interventions with the parents and families has increased generally,[12] and CHIP has highlighted that with CHD at least, such an interest is likely to be productive.

However, there remains much work to do. First, fathers have featured in an ad hoc, almost incidental, way in both the empirical literature which informs such interventions and in the intervention protocols themselves. Part of this relates to the difficulty in engaging fathers or simply engaging *both* parents when there are other competing family demands to be managed to allow participation. However, future research needs to find a way to hear the voices of fathers, and assess their experience of and specific influence on subsequent interventions, as that work which has been done (see chapter: A Family Affair) suggests something distinctive from maternal experiences may be occurring. Secondly, although important, effects sizes, range of impact, and maintenance of gains have been relatively modest or equivocal both in CHIP and in related programs with other chronic illnesses.[12] We need to look at longer-term outcomes, consider the impact of "booster" sessions, and explore pathways of influence and the timing of benefit from secondary prevention interventions. This will require much larger, well-resourced, multicenter trials to rigorously unpack these processes, but findings from studies such as CHIP suggest such investment will be well founded. Thirdly, and related, only larger studies will allow us to vary the various elements of a program such as CHIP to determine if specific elements of CHIP are particularly important or if it is the combined interweave of interventions which is important. Fourthly, CHIP has primarily involved interventions with parents of children with CHD. It may be possible to augment treatment effects by having a specific CHIP component with the children themselves. Camp-based interventions to promote independence and resilience, for example, have shown promise for children with other chronic illnesses and disabilities.[26] Moreover, our own work to be presented in the next chapter highlights how specific interventions with adolescents with CHD themselves can promote improved activity levels,[27] and we have shown how parallel parent and child groups, underpinned by the problem prevention therapy formulated in CHIP can promote improved adjustment in siblings of children with cancer.[28] Thus, combining parent and child interventions within a revised CHIP protocol would appear an exciting avenue of study in the future. Finally, targeting barriers and other translational challenges to implementing programs such as CHIP in different international healthcare contexts needs to be considered, and these considerations are featured in the concluding chapter to this book.

References

1. Visconti KJ, Saudino KJ, Rappaport LA, Newburger J, Bellinger DC. Influence of parental stress and social support on the behavioral adjustment of children with transposition of the great arteries. *J Dev Behav Pediatr*. 2002;23:314–321.
2. Drotar D. *Psychological Interventions in Childhood Chronic Illness*. Washington, DC: American Psychology Association; 2006.
3. Utens EM, Levert E. Psychological aspects in children and adolescents with congenital heart disease and their parents. In: Callus E, Quadri E, eds. *Clinical Psychology and Congenital Heart Disease: Lifelong Psychological Aspects and Interventions*. Italia: Springer-Verlag; 2015.
4. McCusker CG, Doherty N, Molloy B, et al. A randomized controlled trial of interventions to promote adjustment in children with congenital heart disease entering school and their families. *J Pediatr Psychol*. 2012;37:1089–1103.
5. Palermo TM. Evidence-based interventions in pediatric psychology: progress over the decades. *J Pediatr Psychol*. 2014;39:752–753.
6. McCusker CG, Doherty NN, Molloy B, et al. Determinants of neuropsychological and behavioural outcomes in early childhood survivors of congenital heart disease. *Arch Dis Child*. 2007;92:137–141.
7. Casey FA, Stewart M, McCusker CG, et al. Examination of the physical and psychosocial determinants of health behaviour in 4–5 year old children with congenital heart disease. *Cardiol Young*. 2010;20:532–537.
8. Kreuter M, Holt C. How do people process health information? Applications in an age of individualized communication. *Curr Direct Psychol Sci*. 2001;10:206–209.
9. Moola F, Fusco C, Kirsh J. The perceptions of caregivers toward physical activity and health in youth with congenital heart disease. *Qual Health Res*. 2011;21:278–291.
10. Stehl M, Kazak A, Alderfer M, et al. Conducting a randomised clinical trial of a psychological intervention for parents/caregivers of children with cancer shortly after diagnosis. *J Pediatr Psychol*. 2009;34:803–816.
11. Wysocki T, Harris MA, Buckloh LM, et al. Effects of behavioral family systems therapy for diabetes on adolescents' family relationships, treatment adherence and metabolic control. *J Pediatr Psychol*. 2006;31:928–938.
12. Law EF, Fisher E, Fales J, Noel M, Eccleston C. Systematic review and meta analysis of parent and family-based interventions for children and adolescents with chronic medical conditions. *J Pediatr Psychol*. 2014;39:866–886.
13. De Vet K, Ireys H. Psychometric properties of the maternal worry scale for children with chronic illness. *J Pediatr Psychol*. 1998;23(4):257–266.
14. Derogatis IR. *Brief Symptom Inventory: Administration, Scoring and Procedures Manual*. Minneapolis, MN, USA: National Computer Systems; 1993.
15. Cohen JW. *Statistical Power Analysis for the Behavioural Sciences*. 2nd ed. Hillsdale, NJ: Lawrence Erlbaum Associates; 1988.
16. Stein R, Reissman C. The development of an impact-on-family scale: preliminary findings. *Med Care*. 1980;18:465–472.
17. Achenbach TM, Rescorla LA. *Manual for the ASEBA School-Age Forms and Profiles*. Burlington, VT: University of Vermont, Research Center for Children, Youth and Families; 2001.
18. Le Roy S, Elixson EM, O'Brien P, et al. Recommendations for preparing children and adolescents for invasive cardiac procedures: a statement from the American Heart Association Pediatric Nursing Subcommittee of the Council on Cardiovascular Nursing in collaboration with the Council on Cardiolvascular Diseases of the Young. *Circulation*. 2003;108:2550–2564.
19. Marino BS, Lipkin PH, Newburger JW, et al. American Heart Association Congenital Heart Defects Committee, Council on Cardiovascular Disease in the Young, Council on Cardiovascular Nursing, and Stroke Council. Neurodevelopmental outcomes in children with congenital heart disease: evaluation and management: a scientific statement from the American Heart Association. *Circulation*. August 28, 2012;126(9):1143–1172.

20. Institute of Medicine. *Cancer Care for the Whole Patient: Meeting Psychosocial Health Needs.* Washington, DC: The National Academies Press; 2007.
21. National Institute for Health and Clinical Excellence. *Improving Outcomes for Children and Young People with Cancer.* UK: NICE; 2005.
22. Wade S, Walz N, Carey J, et al. A randomized trial of teen online problem solving efficacy in improving caregiver outcomes after brain injury. *Health Psychol.* 2012;31:767–776.
23. Beale IL. Scholarly literature review: efficacy of psychological interventions for pediatric chronic illnesses. *J Pediatr Psychol.* 2006;31:437–451.
24. Fedele DA, Hullmann SE, Chaffin M, et al. Impact of a parent-based interdisciplinary intervention for mothers on adjustment in children newly diagnosed with cancer. *J Pediatr Psychol.* 2013;38:531–540.
25. Sahler OJ, Fairclough DL, Phipps S, et al. Using problem solving skills training to reduce negative affectivity in mothers of children with newly diagnosed cancer: report of a multisite randomized trial. *J Consult Clin Psychol.* 2005;73:272–283.
26. O'Mahar K, Holmbeck G, Jandasek B, Zukerman J. A camp-based intervention targeting independence among individuals with spina bifida. *J Pediatr Psychol.* 2009;35:848–856.
27. Morrison ML, Sands AJ, McCusker CG, et al. Exercise training improves activity in adolescents with congenital heart disease. *Heart.* 2013;99:1122–1128.
28. Besani C, Higgins A, McCusker CG, McCarthy A. A new one day systemic intervention for siblings of children who have cancer and their parents: a pilot and feasibility study. In: *Paper Presented at the 4th Meeting of the International Society for Pediatric Oncology.* Toronto, Canada; 2013.

Healthy Teenagers and Adults: An Activity Intervention

M.L. Morrison, F. Casey

The Royal Belfast Hospital for Sick Children, Belfast, Northern Ireland

BACKGROUND TO EXERCISE TRAINING IN CONGENITAL HEART DISEASE

Improved survival among children with congenital heart disease (CHD) has dramatically increased the number of adolescents and adults with complex heart conditions. As described in chapter "Congenital Heart Disease: The Evolution of Diagnosis, Treatments, and Outcomes," along with the increase in number of CHD patients surviving overall, there is also an increase in the survival of patients with complex lesions.[1] The spectrum of adult congenital heart disease (ACHD) is one of a young population with a relatively low mortality but a high morbidity.[2] Focus has now shifted to this group of patients having the best possible outcome in terms of the quality of life and functional status.

The EuroHeart Survey, a European registry of ACHD patients suggests that exercise intolerance affects around one-third of CHD patients.

163

This includes patients with simple as well as complex lesions.[2] Many patients demonstrate reduced peak oxygen uptake, decreased oxygen saturations, reduced ventilatory anaerobic threshold, and impaired chronotropic response during exercise testing. There is firm evidence that diminished exercise capacity is linked to hospitalization and death in patients with a wide range of conditions including transposition of the great arteries, tetralogy of Fallot, and single ventricle physiology.[3–7] There is also a correlation between poorer exercise tolerance and increasing New York Heart Association (NYHA) class in ACHD patients.[8]

Assessment of exercise ability is utilized to gauge the need for surgical intervention or assess response to medication/treatment. Exercise testing may also be used to confirm and evaluate arrhythmias in patients at risk or those with a suggestive history.[9,10] Actual physical capacity for exercise in CHD patients is variable. Some individuals can exercise normally but others are quite limited. Between these lie the majority of patients for whom exercise recommendation must take into consideration a number of factors including individual motivation and type of exercise proposed.[11]

Historically there has been a lack of specific recommendation from international working groups on this issue, with most guidance comprising summaries of personal recommendations from individual units.[11,12] In 2013, the American Heart Association produced a consensus statement on promotion of physical activity in children and adults with CHD which recommends that counseling to encourage daily participation in appropriate physical activity should form a core part of every patient encounter.[13] This document also recommends that exercise testing be employed prior to engaging in recreational activity to reassure the patient and their family about their ability to be active.

Attitudes to exercise are variable among CHD patients. Some may have other comorbidities, such as neurodevelopmental or orthopedic problems, which make physical activity difficult. They may be deterred by body image issues due to scars, scoliosis, or cyanosis. Some patients have been "overprotected" by parents, teachers, or even health-care professionals predisposing them to exercise intolerance.[14] A number of studies have focused on feelings toward exercise and advice given to CHD patients and their families. Worryingly, these report that discussion about physical fitness and exercise prescription is rare at clinic consultation. Many patients are unaware of what level of activity is suitable for them.[15,16] More encouragingly some studies show positivity in terms of a willingness to participate among patients.[8,16]

The role of exercise-based rehabilitation is well established as the standard of care for patients with acquired cardiovascular disease. The concept of exercise training as a method of improving exercise tolerance in CHD patients has been around since the 1990s.[4] However, until 2014 in the United Kingdom no formal program of exercise rehabilitation existed for patients with CHD.[17] It is only in recent years that, following techniques employed for heart failure patients, the benefits of exercise training programs have begun to extrapolate to CHD as a method of altering morbidity and mortality for the future.

There are a limited number of studies examining exercise training in young people with CHD. These have offered encouraging results with the majority having demonstrated improvements in physical activity along with improvements in exercise performance in the short term. However, the methodologies employed are mixed and some studies include only very small patient numbers. To date studies of pediatric cardiac rehabilitation vary greatly in inclusion criteria, attrition rate, follow-up, and in particular with regard to the structure of the exercise-training program employed or method of activity assessment.[8,18–21] Many require patients to travel to participate in each session. Some papers do acknowledge the difficulty in initiating a lifestyle change but do not approach this in their methodology.[22] Few of the published studies include education, activity counseling, or psychological intervention, all of which are key components of adult cardiac rehabilitation theory.

A key principle noted in this book (see chapters: "The Congenital Heart Disease Intervention Program (CHIP) and Interventions in Infancy" and "Growing Up – Interventions in Childhood and CHIP-School") is that secondary prevention interventions in pediatrics should be timed to occur at key developmental transitions. The developmental crucible of adolescence, during which adult behaviors are tested and become established, is an ideal time for such health promotion strategies, which could potentially influence the public health burden of tomorrow's adults. The unique features of the intervention described here were that health behavior change strategies—motivational interviewing techniques[23]—were used to help lay the foundations for engagement in a personalized exercise plan.[24]

MOTIVATING AND STIMULATING INCREASED EXERCISE ACTIVITY IN ADOLESCENTS WITH CHD: THE BELFAST PROGRAM

From 2008 to 2010 a prospective randomized controlled trial was conducted in Belfast to ascertain if an exercise-based motivational session, followed by an individualized, structured, program of exercise training, could be used to increase physical activity and improve psychological well-being in a group of adolescents with CHD.

One hundred and forty-three patients were recruited. The population consisted of patients with both cyanotic and acyanotic conditions, including those who had both corrective and palliative surgeries and patients with untreated minor lesions. As in previous research patients with syndromic diagnosis, major learning difficulty, or those for whom exercise was contraindicated were excluded. Participants were randomized to intervention or control groups using balanced blocks. Table 11.1A summarizes the characteristics of both groups at baseline. There were no significant differences between the groups in terms of age, sex, growth parameters, or diagnostic category. There was a significant difference in socioeconomic deprivation index; therefore, this was used as a covariate in subsequent analysis.

For all patients, peak exercise capacity was assessed using a formal exercise stress test on a treadmill ergometer; participants were encouraged to exercise to their maximum capacity. Day-to-day, free-living activity was measured with an Actigraph GT1M accelerometer. Amount of time spent in moderate to vigorous physical activity (MVPA) was estimated using activity counts and compared to current UK guidelines which suggest that young people should get more than 60 min of daily physical activity.[8]

Seventy-two patients attended a one-day workshop (in groups of 12) which included the following elements:

- Psycho-education with respect to exercise in physical health generally and following CHD in particular.
- Completion of a stage of change questionnaire in which the importance of exercise, together with confidence and readiness to change were rated on a 10-point Likert scale.
- Group generation of problems and obstacles to engagement with regular exercise.
- Problem-solving therapy (see chapters: "The Congenital Heart Disease Intervention Program (CHIP) and Interventions in Infancy" and "Growing Up – Interventions in Childhood and CHIP-School") overview and practice in small groups.
- Motivational interviewing interventions by program facilitators (eg, empathy, promoting dissonance to motivate change, visualization of lifestyle, and routine with exercise-woven throughout).[23]
- Each participant was seen individually and given a written, personalized exercise training plan containing suggestions for how to increase their activity at home over the next 6 months, in a manner suitable for their diagnosis.

The activity day was followed up with a personalized fact sheet which also summarized the psycho-education, specific discussions, problems, and solutions generated in that group. Each participant was contacted once a month to check on their progress and discuss any problems. No major difficulties with the program were identified during these conversations; intervention group participants found them to be encouraging and supportive.

TABLE 11.1 (A) Characteristics of Intervention and Control Groups at Baseline. (B) Exercise Parameters of Intervention and Control Groups at Reassessment

	Intervention Group ($n=72$)	Control Group ($n=71$)	p Value
(A)			
Mean age (years)	15.24	15.89	0.06
Number of males	48 (66.7%)	38 (53.5%)	0.11
Diagnostic group			0.09
Minor CHD	19 (26.4%)	20 (28.1%)	
Acyanotic corrected	27 (37.5%)	34 (47.9%)	
Cyanotic corrected	15 (20.8%)	15 (21.2%)	
Cyanotic palliated	11 (15.3%)	2 (2.8%)	
Major CHD	53 (73.6%)	52 (73.2%)	0.96
BMI SDS (mean)	−0.06	0.26	0.12
Height SDS (mean)	−0.23	−0.12	0.57
Weight SDS (mean)	−0.18	0.13	0.99
Deprivation	348.7	258.9	<0.01
Activity per week			0.37
None	3 (4.2%)	6 (8.5%)	
Once	7 (9.7%)	12 (16.9%)	
Twice	17 (23.6%)	10 (14.1%)	
3 times	10 (13.9%)	10 (14.1%)	
More than 3 times	35 (48.6%)	33 (46.5%)	
(B)			
Duration of EST (mean ± SD)			$F(1, 97)=5.3$ Pillai's trace = 0.05 Partial $\eta^2=0.05$ $p<0.05$
Baseline	10.9 ± 3.2	12.1 ± 3.5	
Reassessment	12.0 ± 3.8	12.02 ± 3.7	
Predicted VO$_{2max}$ (mean ± SD)			$F(1, 96)=6.6$ Pillai's trace = 0.06 Partial $\eta^2=0.06$ $p<0.05$
Baseline	35.0 ± 7.4	37.8 ± 8.6	
Reassessment	37.4 ± 8.8	37.5 ± 8.6	
Maximum METs achieved (mean ± SD)			$F(1, 97)=2.7$ Pillai's trace = 0.03 Partial $\eta^2=0.03$ $P=0.10$
Baseline	12.9 ± 3.5	14.3 ± 3.9	
Reassessment	15.6 ± 2.2	14.0 ± 4.3	
Average minutes of MVPA per day (mean ± SD)			$F(1, 88)=81.3$ Pillai's trace = 0.48 Partial $\eta^2=0.48$ $P<0.001$
Baseline	28.4 ± 20.1	32.7 ± 28.7	
Reassessment	57.2 ± 32.2	29.2 ± 27.3	

Outcomes were encouraging. Post-intervention session means for the motivational rating scale components were considerably higher than those obtained pre-session. Table 11.2 lists pre- and post-session means along with significant p values, and confidence intervals for domains of confidence to change, importance of change, and readiness to implement change, highlighting the positive effect the session had on these issues.

There were distinct differences on exercise testing between intervention and control groups at reassessment. Duration of exercise test increased by 1 min 5 s for the intervention group (p value <0.05) along with a significant difference in predicted VO_{2max} (p value <0.05). For free-living activity there was a significant increase in average minutes of MVPA per day for the intervention group from baseline to reassessment (p value <0.001) while minutes of MVPA remained much the same for the control group (Table 11.1B and Fig. 11.1). At baseline 14 patients met the current UK recommendation for more than 60 min MVPA per day. This doubled to 29 participants at reassessment. There were no adverse effects reported by participants and no mortality during the intervention program.

COMMENT AND CONCLUSIONS

The Belfast study demonstrated that the majority of patients with CHD participated in exercise each week and a large number of them played sport competitively as evidenced by completed activity questionnaires. This included the most complex patient group. There were no significant differences in activity score at baseline among the diagnostic groups indicating that patients with complex CHD consider themselves to be as active day to day as those with less complex problems. However, subjective measures of activity are inherently flawed and have been shown to be unreliable.[3] Exercise testing in this group confirmed that complex patients are limited at peak exercise, even if they consider themselves to be asymptomatic. These findings are similar to those reported by other authors.[3]

As the patient group is young and relatively "well" it may be that they have not yet reached a point in the natural history of their condition where

TABLE 11.2   Pre- and Post-session Means for Motivational Ratings Scale Parameters

	Pre-session	Post-session	p Value (95% Confidence Interval)
Importance	6.84 ± 2.36	7.79 ± 2.28	<0.01 (−1.27, −0.65)
Confidence	6.65 ± 2.19	7.62 ± 2.15	<0.01 (−1.35, −0.59)
Readiness	6.75 ± 1.99	7.69 ± 1.96	<0.01 (−1.27, −0.61)

exercise limitation truly becomes apparent. Low levels of activity documented in ACHD patients may reflect a poor exercise capacity, development of more symptoms, or a general deterioration in health with age.[3,25] Studies in ACHD patients also show that there is potential for increasing exercise capacity through training[26] indicating that this phenomenon cannot be entirely blamed on symptomatology alone. Perceived functional status and self-efficacy are important determinants of activity, meaning

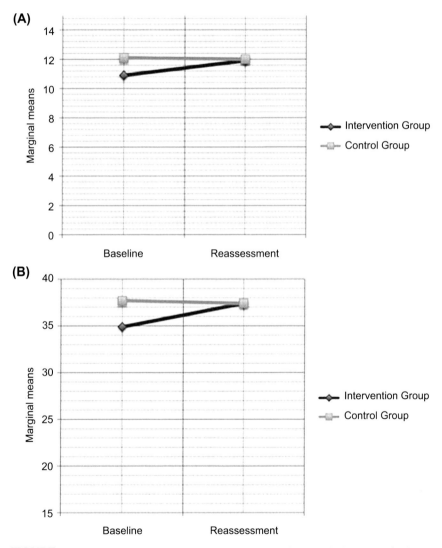

FIGURE 11.1   Plots of marginal means for (A) duration of EST and (B) estimated VO_{2max} (covariate—deprivation).

that it is not what someone is capable of that is important but what they perceive they are capable of.[27]

The Belfast study is unique in terms of exercise training intervention in CHD. It utilized health behavior change strategies based on motivational interviewing techniques to educate about exercise and stimulate and maintain change in exercise behavior. Traditional exercise programs are difficult to maintain. They require significant input in terms of time, facilities, personnel, and resources. Few include educational or psychological intervention components. In the AHA Scientific Statement Longmuir et al.[13] emphasize the importance of effective behavior change counseling as a means to identify and implement plans to achieve personally relevant goals with regard to exercise behavior.[13]

In this program participants received an individual training plan to implement at home. During the discussion sessions there was opportunity to identify and address patient-perceived obstacles, problems, and solutions. Education and personal choice form key components of the exercise plans; such elements are likely to increase adherence and achieve sustained behavior change.[13] The approach is patient centered and is successful in moving the individual through "stages of change" with regard to exercise behavior. Motivational ratings scales completed pre- and post-session indicated the very strong impact this problem-solving and motivational interviewing intervention had on confidence and readiness to change exercise behavior.

The Belfast intervention program takes a multilevel approach targeting the patient and their lifestyle. It can be implemented with fewer resources than traditional exercise training programs and on a sustainable basis. The program can easily be adapted to include patients outside urban areas and has been shown to have a significant impact on activity levels. At reassessment the intervention group had significantly increased both their peak exercise capacity and their day-to-day MVPA. The number reaching the recommended 60 min of MVPA doubled.

There is evidence that most CHD patients are willing to participate in exercise but are uncertain as to the safety or benefits. Reviews advise that activity participation should form part of outpatient clinic discussion.[8,15] A motivational style intervention and a home-based exercise training plan would fit neatly into formal transition programs for teenagers with CHD.

The Belfast study confirms that adolescents with CHD can exercise safely and that such training programs are both feasible and beneficial for young people with CHD. This study is unique in that it employed psychological methods to maximize the impact of its intervention. Consideration for issues such as maintenance of activity, psychological well-being, and promotion of good lifestyle choices must be given when planning formal services for this group of patients in the future.

References

1. Warnes CA. The adult with congenital heart disease: born to be bad? *J Am Coll Cardiol.* 2005;46:1–8.
2. Engelfriet P, Boersma E, Oechslin E, et al. The spectrum of adult congenital heart disease in Europe: morbidity and mortality in a 5 year follow-up period. *Eur Heart J.* 2005;26(21):2325–2333.
3. Diller GP, Dimopoulos K, Okonko D, et al. Exercise intolerance in adult congenital heart disease: comparative severity, correlates, and prognostic implication. *Circulation.* August 9, 2005;112(6):828–835.
4. Longmuir PE, Tremblay MS, Goode RC. Postoperative exercise training develops normal levels of physical activity in a group of children following cardiac surgery. *Pediatr Cardiol.* July 1990;11(3):126–130.
5. Fernandes SM, McElhinney DB, Khairy P, Graham DA, Landzberg MJ, Rhodes J. Serial cardiopulmonary exercise testing in patients with previous Fontan surgery. *Pediatr Cardiol.* 2010;31:175–180.
6. Giardini A, Specchia S, Tacy TA, et al. Usefulness of cardiopulmonary exercise to predict long-term prognosis in adults with repaired tetralogy of Fallot. *Am J Cardiol.* 2007;99:1462–1467.
7. Giardini A, Hager A, Lammers AE, et al. Ventilatory efficiency and aerobic capacity predict event-free survival in adults with atrial repair for complete transposition of the great arteries. *J Am Coll Cardiol.* 2009;53:1548–1555.
8. Dua JS, Cooper AR, Fox KR, Stuart GA. Physical activity in adults with congenital heart disease. *Eur J Cardiovasc Prev Rehabil.* 2007;14(2):287–293.
9. Arena R, Meyers J, Williams MA, et al. Assessment of functional capacity in clinical and research setting. A scientific statement from the American Heart Association committee on exercise, rehabilitation, and prevention. *Circulation.* 2007;116:329–343.
10. Tomassoni T. Role of exercise in the management of cardiovascular disease in children and youth. *Med Sci Sports Exerc.* 1996;28:406–413.
11. Cullen S, Celemajer D, Deanfield J. Exercise in congenital heart disease. *Cardiol Young.* 1991;1:129–135.
12. Sklansky MS, Bricker JT. Guidelines for exercise and sports participation in children and adolescents with CHD. *Prog Pediatr Cardiol.* 1993;2(3):55–66.
13. Longmuir PE, Brothers JA, de Ferranti S, et al. Promotion of physical activity for children and adults with congenital heart disease. A scientific statement from the American Heart Association. *Circulation.* 2013;127:2147–2159.
14. Tikkanen AU, Oyaga AR, Riano OA, et al. Paediatric cardiac rehabilitation in congenital heart disease: a systematic review. *Cardiol Young.* 2012;22:241–250.
15. Swan L, Hillis WS. Exercise prescription in ACHD: a long way to go. *Heart.* 2000;83:685–687.
16. Kendall L, Parsons JM, Sloper P, Lewin RJ. A simple screening method for determining knowledge of the appropriate levels of activity and risk behaviour in young people with congenital cardiac conditions. *Cardiol Young.* 2007;17:151–157.
17. Pieles GE, Horn R, Williams CA, Stuart AG. Paediatric exercise training in prevention and treatment. *Arch Dis Child.* 2014;99:380–385.
18. Fredriksen PM, Kahrs N, Blaasvaer S, et al. Effect of physical training in children and adolescents with congenital heart disease. *Cardiol Young.* 2000;10(2):107–114.
19. Rhodes J, Curran T, Camil L, et al. Impact of cardiac rehabilitation on the exercise function of children with serious congenital heart disease. *Pediatrics.* 2005;116:1339–1345.
20. Rhodes J, Curran T, Camil L, et al. Sustained effects of cardiac rehabilitation in children with serious congenital heart disease. *Pediatrics.* 2006;118:586–593.
21. Thaulow E, Fredriksen PM. Exercise and training in adults with congenital heart disease. *Int J Cardiol.* 2004;97(Suppl 1):35–38.

22. Longmuir PE, Tyrrell PN, Corey M, et al. Home-based rehabilitation enhances daily physical activity and motor skill in children who have undergone the Fontan procedure. *Pediatr Cardiol*. 2013;34:1130–1151.
23. Rollnick S, Mason P, Butler C. *Health Behavior Change*. London: Elsevier; 2007.
24. Morrison ML, Sands AJ, McCusker CG, et al. Exercise training improves activity in adolescents with congenital heart disease. *Heart*. 2013;99:1122–1128.
25. Dimopoulos K, Diller GP, Piepoli M, et al. Exercise intolerance in adults with CHD. *Cardiol Clin*. 2006;24:641–660.
26. Dua JS, Cooper AR, Fox KR, Stuart GA. Exercise training in adults with CHD: feasibility and benefits. *Int J Cardiol*. 2010;138:196–205.
27. Bar-Mor G, Bar-Tal Y, Krulik T, Zeevi B. Self-efficacy and physical activity in adolescents with trivial, mild or moderate congenital cardiac malformations. *Cardiol Young*. 2000;10:561–566.

12

Conclusions and Future Directions for Neurodevelopmental Research and Interventions in Congenital Heart Disease

C. McCusker

The Queens University of Belfast, Belfast, Northern Ireland; The Royal Belfast Hospital for Sick Children, Belfast, Northern Ireland

AN EVOLUTION IN UNDERSTANDING

As noted at the outset, this book has been about constructing an evidence-based narrative which helps us better understand the outcomes, the processes behind the outcomes, and potential interventions to improve the outcomes, for children with congenital heart disease (CHD) and their families.

While much of our work has been focused on preventing negative outcomes, it is important to start this final chapter with a reminder that for most children with CHD, and their families, outcomes are positive. Not only have survival rates for significant CHD increased from about 20% in the 1950s to 90% in 2006,[1] but many studies suggest these survivors are as well adjusted, or sometimes more adjusted, as their healthy peers.[2,3] Doherty and Utens in chapter "A Family Affair" and Katz and her colleagues in chapter "The Adult with Congenital Heart Disease" review evidence which suggests the same is true for the parents and siblings of these children and for adults with CHD. Utens has referred to this as *post-traumatic growth* following the trauma of CHD,[4] and other work has suggested that the very processes which can amplify risk for maladjustment also confer resilience over time.[5]

However, such positive outcomes are not enjoyed by all, with significant numbers evidencing difficulties in neurodevelopment, psychological and social functioning, and educational and occupational attainments. Throughout the chapters in this book it has been clearly discussed that risk for psychosocial maladjustment is probably 2–3 times higher in this population than that in healthy peers. Risk also appears elevated in parents and siblings. Moreover, it has been a striking observation that despite surgical advances in the past couple of decades, rates of psychosocial difficulties have remained stubbornly unchanged.[6] Similar focus now needs to be given to developing psychological understanding and interventions if we are to improve these outcomes for this growing population of child and adult survivors.

Such improved understanding has been emerging. This volume has brought together the work of leading researchers in the field from across Europe and North America. As we noted in chapter "Is There a Behavioral Phenotype for Children with Congenital Heart Disease?," there has been much "noise" in the research literature, often created by the problems which have bedeviled pediatric psychology research more generally.[7] However, consistent stories are starting to emerge and underpinned by improvements in research design.

Thus, conclusions reported in this book are bolstered by longitudinal research designs which have tracked the same cohorts across time and can better discriminate cause and effect processes. Together we have been able to incorporate children with multiple CHD diagnoses which have allowed us to better understand how factors such as cyanosis, open versus closed surgical interventions, and having a corrected versus palliated condition might impact on outcome. We have been able to incorporate more stringent control groups (eg, siblings and those with mild defects as well as healthy peers) and indeed have moved beyond consensus statements with respect to interventions, to evaluating such under randomized controlled trial conditions and with longer-term follow-ups than traditionally reported. We have paid more careful attention to who provides data for interpretation and the potential noise this creates (eg, parents vs. teachers and self-reports). Moreover we have attended to the potentially confounding influence of using popular measures such as the *Child Behavior Checklist* (CBCL) where "somatic" subscales, based on mental health presentations, risk distorting psychological profiles for pediatric populations. What then have been the stories which have emerged?

A NEURODEVELOPMENTAL PHENOTYPE: MOTOR DIFFICULTIES TO EXECUTIVE FUNCTIONING

If we exclude those children with comorbid developmental syndromes, work has generally found that these children fall in the "average" range of intellectual ability, albeit toward the lower end of the range on average. However, the work presented in Section II of this book suggests that deficits are evident in specific cognitive processes which do compromise realization of intellectual potential and academic attainments. Most consistently, problems have been reported within the realm of visuomotor abilities (eg, motor development, coordination, sequencing motor movements, and constructional abilities). Indeed for some time visuomotor coordination difficulties appeared a likely candidate for the neurodevelopmental phenotype of this population. However, as noted in chapters "A Longitudinal Study from Infancy to Adolescence of the Neurodevelopmental Phenotype Associated with d-Transposition of the Great Arteries" and "Neurodevelopmental Patterns in Congenital Heart Disease across Childhood: Longitudinal Studies from Europe," this perception was altered and expanded when studies started to extend beyond infancy and early childhood, and toward middle and late childhood and adolescence.

Perspectives across childhood suggest that while visuomotor problems may actually reduce in some, apparently new problems, not discerned at earlier ages, emerge—problems with attention, integrative reasoning and narratives, and executive functioning.[8,9] Whether engaged in perceptual–motor tasks or narrative memory tasks, Bellinger and Newburger noted the propensity for their older childhood sample to get *lost in the detail* (see chapter: A Longitudinal Study from Infancy to Adolescence of the Neurodevelopmental Phenotype Associated with d-Transposition of the Great Arteries) at the expense of structure and organization. Hövels-Gürich and McCusker (in chapter: Neurodevelopmental Patterns in Congenital Heart

Disease across Childhood: Longitudinal Studies from Europe) argue that underlying difficulties with executive functioning may be manifesting themselves in different ways at different stages of development and on associated neuropsychological tasks therein. Thus, before becoming more automatic in later development, visuomotor abilities are controlled by the same executive function processes,[10] which in later development become most apparent in relation to integrative reasoning, planning, organization, and sequencing of conceptual information. The emergent phenotype we are proposing, therefore, is one related to common difficulties with executive functioning. The fact that such difficulties are often associated with problems in social cognition and that such difficulties have also begun to be noted in this population,[11] is consistent with this conclusion.

For some time it was thought that cyanosis and cardiopulmonary bypass were the key risk factors for adverse neuropsychological outcomes. This was chiefly because both can compromise oxygenation of the blood and thus the brain, rendering neurological insults more likely. As noted in Section II these factors remain of significant import, with MRI scans for example showing greater neurological insults in cyanotic compared to acyanotic conditions,[13] and our own work often demonstrating somewhat greater neuropsychological deficits in those with cyanotic conditions. However, our own work, and that of others, has also increasingly challenged this story. Outcomes across diagnostic groups in our Belfast center and elsewhere have suggested that children with acyanotic conditions are just as susceptible to at least some neurocognitive deficits as those with cyanotic deficits.[9,14–16]

Advances in neuroimaging studies are providing some explanation for this in documenting intrauterine delay of brain maturation in CHD more generally. Thus an "encephalopathy" of CHD has been proposed,[17] which confers direct risk for these children but which undoubtedly also makes them more vulnerable to the additional neurological impact of cyanosis and cardiopulmonary bypass.

While protective neurological interventions during surgery are important here, a key message throughout these chapters on neurocognitive outcomes is that family and environmental factors are also important and as such are other important foci for interventions. Our work has highlighted the importance of maternal distress, worry, and parenting abilities also in contributing to some of the variance in neurocognitive outcomes.[9,14,18,19] Thus such familial conditions may set the scene for lack of stimulation, reciprocal social interactions, and reduced opportunities for motor coordination practice in the infant. Benefits for guided neurodevelopmental stimulation shown in our work (see chapter: The Congenital Heart Disease Intervention Program (CHIP) and Interventions in Infancy) support the importance of these factors. Considered together, we arrive in chapter "Neurodevelopmental Patterns in Congenital Heart Disease across

Childhood: Longitudinal Studies from Europe" at a four-factor model for understanding and improving neurodevelopmental outcomes which involves concomitant neurological features, peri- and postoperative management, cyanosis, and family and environmental factors.

PSYCHOLOGICAL ADJUSTMENT: THE IMPORTANCE OF PERSONAL AND INTERPERSONAL COMPETENCIES

If family and environmental factors have a part to play in understanding neurodevelopmental outcomes, then the work in Section III of this book suggests they take center stage when understanding psychological adjustment in CHD. As Rooney notes in chapter "Historical Perspectives in Pediatric Psychology and Congenital Heart Disease" of this book, from the earliest days of pediatric psychology research, a simple linear relationship between disease severity and childhood adjustment has not been evident. While clearly posing specific challenges that cannot be ignored, McCusker and Casey in chapter "Is There a Behavioral Phenotype for Children with Congenital Heart Disease?" present evidence which shows how systemic factors—maternal mental health, parenting, and understanding and appraisals of the condition and cohesiveness of family functioning—repeatedly play a greater part in determining psychological outcomes for these children than the severity of the disease itself.[14,19] Katz and colleagues in chapter "The Adult with Congenital Heart Disease" show that the same pattern continues to apply to adults living with CHD. What problems though do these children present with?

Our longitudinal work at the Belfast center has suggested that the emergent behavioral phenotype is not primarily related to anxiety, depression, or mood disturbance as had sometimes been suggested in earlier work. Although we have argued in chapter "Is There a Behavioral Phenotype for Children with Congenital Heart Disease?" that such mood disturbances may well emerge as secondary consequence, we have been struck by the fact that when we control for the somatic element of popular outcome measures (as noted in the preceding paragraphs) and consider independent teacher, as well as maternal, respondents then a slightly different story emerges. The story suggests that the primary problems manifested by these children relate to personal and interpersonal competencies such as social relationships, physical activity, thought and attentional disturbances, and school functioning.[19] Indeed, as noted by Bellinger and Newburger in chapter "A Longitudinal Study from Infancy to Adolescence of the Neurodevelopmental Phenotype Associated with d-Transposition of the Great Arteries," the neurocognitive deficits in executive functioning and social cognition might be expected to compromise the cognitive competencies required to manage the increasingly complex transactions of child and adolescent relationships, in much the same way as physical limitations (or perceptions thereof) might be expected to compromise physical competencies to engage with—especially in active and effortful contexts. This is consistent with the retrospective accounts of adults with CHD described in chapter "The Adult with Congenital Heart Disease," where difficulties noted related to "feeling different" from others and social integration, rather than primary problems with anxiety or depression. Thus, it is suggested that interventions of secondary prevention, should be about promoting competencies in these children, rather than offering mood management interventions per se.

It has been a salient theme of this book that given that families determine so much of the outcomes for children with CHD, then interventions should be family focused. Doherty and Utens in chapter "A Family Affair" highlight that parents are themselves at risk for adverse psychological sequelae. Again, disease severity is not itself the primary determining factor here but rather appraisals, parenting, and coping competencies and degree of collaboration in medical caregiving.[20] Taken together with our understanding of the processes involved in determining child outcomes, this increasingly points toward interventions which are about building resilience through the promotion of caretaking, parenting, and transactional competencies within the medical setting for parents, and then through parents promoting cognitive development, activity, and interpersonal competencies in the child.

AN INTEGRATIVE MODEL OF UNDERSTANDING

Fig. 12.1 summarizes in diagrammatic form where the evidence base discussed in this book has taken us to, in understanding outcomes for children with CHD and their families. It also highlights where our intervention efforts might best be targeted within different parts of the system. Key points encapsulated in this model include:

- Disease factors (such as cyanosis) and surgical factors (such as length of time on cardiopulmonary bypass) may interact with concomitant neurological features in amplifying personal risk for adverse outcomes; the latter may, however, exert an independent effect and thus children with acyanotic conditions and indeed who have undergone "closed" interventional procedures may also be at risk.
- Family factors are highlighted as amplifying or moderating the risk for adverse neurodevelopmental and psychological outcomes conferred by these biological factors. These are likely to moderate both neurodevelopmental outcomes and psychological adjustment, although the impact on the latter appears greater.

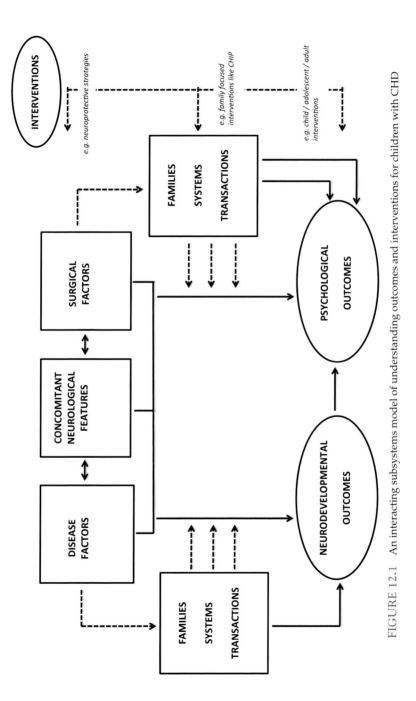

FIGURE 12.1   An interacting subsystems model of understanding outcomes and interventions for children with CHD

- Neurodevelopmental outcomes are themselves likely to exert direct effects on psychological adjustment in terms of compromised cognitive competencies to manage interpersonal contexts including school and peer relationships.
- This model presents a conceptual system which opens up intervention possibilities. Specifically, it has been argued that intervening at the family–systems level offers a significant focus for intervention, the benefits of which have been outlined in Section IV of this book. Other potential points of intervention are noted.

EFFECTIVE INTERVENTIONS

The model outlined above highlights first and foremost the central importance of families and systems for both psychological and neurodevelopmental outcomes. Indeed psychosocial interventions which have been solely focused on the child have seen benefits compromised by anxiety and worry in parents.[20] Thus our *Congenital Heart disease Intervention Program* (CHIP) was based on the premise that by working through and improving outcomes for the family, we could improve outcomes for the child with CHD.

Secondly, the work outlined in Sections II and III suggests that underlying difficulties, shown by these children and indeed their families, are shortfalls in the competencies required to meet the challenge of the disease. Thus, our interventions have not been about mood management per se, although secondary benefits therein were an important index of impact. Rather, interventions have been about promoting competencies in both the parents and the children. Chapters "The Congenital Heart Disease Intervention Program (CHIP) and Interventions in Infancy" and "Growing Up: Interventions in Childhood and CHIP–School" highlight that these competencies related to infant caretaking behaviors, attachment and feeding transactions, careful stimulation of neurocognitive pathways, promoting activity and independence in the pre-school child, facilitating understanding and coping with medical interventions, and actively attending to the healthy functioning of the whole family system. Embedding specific interventions within the framework of problem-solving therapy not only facilitated generalizability of gains, but further engendered development of personal competencies therein.

Thirdly, while we were not aiming at mood management interventions per se, it was clear to us that distress, confusion, and lack of understanding gets in the way of the adaptation and adjustment strategies we were keen to promote. Thus, at the outset both CHIP programs (*CHIP–Infant* and *CHIP–School*), in different ways (through groups or the use of a DVD highlighting the narratives of "experienced" parents), recognized the importance of emotional processing and meaning making, and incorporated narrative processes to facilitate shifts therein.

Finally, the importance of understanding and empowering collaboration in their child's health care had been seen to be associated with parental adjustment (see chapter: A Family Affair). Thus both programs used different age- and stage-specific strategies to increase collaboration and understanding, not just within the specialist unit but also in outreach to community health and education systems which would be instrumental to the subsequent care and development of the child.

The findings reported have been very encouraging. In *CHIP–Infant*,[17] *CHIP–School*,[21] and our adolescent activity intervention,[22] we saw, under controlled trial conditions, improvements in maternal mental health and family functioning and neurodevelopmental gains in the infants, behavioral adjustment gains in the children, and improved activity and exercise in the teenagers. We also found secondary gains in terms of feeding behavior in infancy and reduced "sickness" behavior and school absences in the child.

In sum, we have provided a model which translates the accumulating knowledge base on outcomes for children with CHD and their families, into an effective intervention framework and one that can be adapted to key developmental transitions. A core feature of this intervention is the problem-solving therapy which in a 2014 meta-analysis was shown to be the intervention that currently commands the greatest effect size changes generally in the pediatric psychology literature.[23] However, the other elements of the CHIP program—meaning making, specific foci related to developmental transition, and interventions with the caregiving systems surrounding the family have arguably resulted in a wider range and depth of gains than has been reported in interventions with pediatric populations more generally.[7] Where next then for such work in CHD?

FUTURE DIRECTIONS AND TRANSLATIONAL POTENTIAL

We have begun to get a clearer picture of the neurodevelopmental phenotype in CHD. However, our current tools for clinical neuropsychological assessment need to be developed to better explore the processes behind the deficits. Thus, for example, 2013 experimental analyses of performance on executive functioning tasks suggested that a "sluggish cognitive tempo" may underpin performance in such tasks in children with CHD, distinguishing them from similar cognitive "outputs" demonstrated in ADHD.[24] This will be important in exercising caution in making comorbid diagnoses such as ADHD, and indeed autistic spectrum disorders, based on neurodevelopmental features which can look similar, but have different etiologies and treatment implications. Future work in delineating these neuropsychological processes will benefit from greater synergy with neuroimaging research. Advances in the latter are promising—both in terms of identifying the organic basis to neurodevelopmental problems and in terms of early identification of risk.

A model of understanding both neurodevelopmental and psychological outcomes has been put forward in chapters "Neurodevelopmental Patterns in Congenital Heart Disease across Childhood: Longitudinal Studies from Europe" and "Is There a Behavioral Phenotype for Children with Congenital Heart Disease?" of this book. Larger multicenter trials are required to further test the percentage variance explained by each of the components involved to further focus the intervention efforts. However, perhaps of even greater importance will be reviewing the conceptual bases to adjustment models in children with CHD, and indeed chronic illness and disability more generally.

There may be much to be gained from shifting our focus away from psychopathology, and the associated problems with measurement tools which have been discussed earlier, and toward the concept of resilience. The qualitative research with siblings, fathers, and adults discussed in chapters "A Family Affair" and "The Adult with Congenital Heart Disease" highlights that outcomes in relation to identity, role coherence, and balancing coping behaviors with emotional distress at the same time, may be relevant here and may be closer to capturing outcome trajectories than simply assessing mood disturbance at some point in time after CHD impacts. Such research may be important in delivering theoretical frameworks which can accommodate the often contradictory findings in the pediatric psychology literature such as maladjustment versus posttraumatic growth. Risk and resilience may perhaps be two sides of the same coin, but current models of understanding are some way behind this. Next steps will be about translating such concepts into measures which can be tested for generalizability in greater numbers. Related, current outcome measures more generally need to evolve beyond those rooted in mental health traditions and greater efforts need to be made toward ensuring that fathers and siblings are better represented in our evidence base pertaining to both family outcomes and indeed interventions.

We have seen evidence in this book that our evolving conceptual models can indeed translate into effective intervention practices. Further trials are required to answer specific questions. For example, it will be important to know if specific elements of a program like CHIP are key (ie, the problem-solving therapy, the narrative therapy, the specific parent training, and the wider systemic interventions) or whether the combined elements are necessary for impact. Multicenter trials will be required to answer this question. Furthermore, it will be important to determine whether such interventions have cross-cultural generalizability, and whether extending the program with "booster" sessions across development can amplify gains. Given that these interventions are essentially preventive in nature, longer-term outcome points, beyond 6 months or a year, may be required as there is likely to be a lag between parenting changes, for example, and behavioral outcomes for the child.

We need to look at supplementing the family-focused interventions described in this book with the child-specific interventions indicated in Fig. 12.1, with respect to neurodevelopment and psychological adjustment. While routine monitoring of neurodevelopmental outcomes has been called for by the *American Heart Federation*,[25] and is likely to be important, this will be of limited utility if we do not have effective school-based interventions available for those who we identify as requiring it. Finally, we need to look at translating the findings presented in chapter "The Adult with Congenital Heart Disease" of this book into effective interventions for those living with CHD in adulthood.

Advancing such intervention programs of research will be challenging, and translating the findings of such into routine clinical practice will be still more challenging. There are many practical hurdles to overcome in facilitating access to such intervention programs in families who may have little time or energy left over from managing the multiple demands of medical interventions. The efforts by Stehl and colleagues, for example, in bringing forward their problem-solving intervention in childhood cancer to the diagnostic stage, but then encountering low participation rates exemplifies this challenge.[26] We need to think of creative ways to improve feasibility as much as effectiveness. In *CHIP–Infant* we incorporated such interventions into the inpatient setting, where the program offered something practical and helpful to do during the many long hours of simply being around for their child. Participation rates were consequently very high. Wade and colleagues have shown how similar problem-solving interventions for teenagers with acquired brain injuries may be effectively delivered online.[27] This offers a further rich focus for future research in the CHD population and also one which may minimize translational cost challenges.

Finally, the health-care economics of such programs will be an important focus for future research. Health-care systems across the world, whether funded through private insurance schemes such as in North America, or by the sort of nationalized health service seen in the United Kingdom, are facing unprecedented challenges. These have been created by the confluence of an increasingly aging population, increased survival rates from diseases (such as CHD) which would have been fatal some years ago, increased options for new health-care interventions (and population awareness of and demand for such), and at the same time limited, or even shrinking, financial resources to fund these. We need therefore to build into programs such as CHIP outcomes in relation to costs and cost savings. This will be unfamiliar territory for many researchers in the field. In CHIP we were able to look at reductions in medical consultations and reduced absenteeism from school which partially relate to demonstrating savings within the domains of disability costs and educational disadvantage. However, it would seem likely that the emergent trend toward providing dollar and pound savings will be a future requirement. The improvements in the evidence base for pediatric psychology interventions has gone some way, but needs to go further, toward establishing such interventions as "first-line" treatments.[7] The feasibility of weaving such into the routine frameworks of health-care interventions for children with CHD will require attention to all of these factors.

CONCLUDING COMMENTS

At the outset of this book we considered the personal stories of Lily, Katie, Molly, Malachy, and several other children with CHD who we have known and worked with. Our work has illuminated those stories. We begin to understand better why Lily, with a relatively minor and corrected heart defect experienced more restrictions in physical activity in early childhood than Katie, with a more complex and uncorrected defect. We begin to see how baby Molly's failure to thrive could be better helped by working with her mother's feelings, self-appraisals, and feeding transactions than by simply resigning understanding to the impact of the disease on capacity to feed. We better understand why and how Malachy and Peter might be experiencing school and peer relationship difficulties and the sorts of psycho-educational, remedial, and support strategies which would be of benefit to their parents, teachers, and indeed the children themselves. Finally, we begin to see how the exercise fears, health-related anxieties, and dispositional eccentricities of Grainne, Charlie, and Mary, may have their genesis in specific cognitive traits and family transactions which sought to protect their children from harm and distress.

At the *Royal Belfast Hospital for Sick Children*, where the CHIP program was born, we too have faced the challenge of maintaining our structured intervention programs when research funding ended. However, the end of the funding has not meant the end of the interventions. Our manuals and DVDs continue to serve as a legacy of the program for all the families who pass through our unit. More importantly, the change in the culture of the unit, of the knowledge and skills which are now woven through the work of the pediatric cardiologists, nurses, psychologists, and other professionals in the unit has been enduring. Most fundamentally, we have come to recognize that CHD represents an essential family trauma, but the more we can facilitate meaning making, and a proactive problem-solving stance toward finding the competencies required to manage each developmental transition challenged by the disease, the more likely we are to promote resilient families and the resilient children which will emerge as a consequence.

References

1. Reid G, Webb G, Barzel M, McCrindle B, Irvine J, Siu S. Estimates of life expectancy by adolescents and young adults with congenital heart disease. *J Am Coll Cardiol.* 2006;48:349–355.
2. Utens E, Verslusis-Den Bieman H, Witsenburg M, Bogers AJ, Verhulst FC, Hess J. Cognitive and behavioural and emotional functioning of young children awaiting elective cardiac surgery or catheter intervention. *Cardiol Young.* 2001;11:153–160.
3. Salzer-Mohar U, Herle M, Floquet P, et al. Self-concept in male and female adolescents with congenital heart disease. *Clin Pediatr.* 2002;41:17–24.
4. Utens EM, Versluis-Den Beiman HJ, Witsenburg M, Bogers AJ, Hess J, Verhulst FC. Does age at the time of elective cardiac surgery or catheter intervention in children influence the longitudinal development of psychological distress and styles of coping of parents? *Cardiol Young.* 2002;12(6):524–530.
5. Kennedy L, McCusker CG, Russo K. *Walking the Line Between Risk and Resilience: The Experience of Young People Who Have a Brother or Sister Living with Congenital Heart Disease* [Thesis submitted in partial fulfilment of the requirement of Doctorate in Clinical Psychology]. Queens University Belfast; 2014.
6. Spijkerboer AW, Utens EM, Bogers AJ, Helbing WA, Verhulst FC. A historical comparison of long-term behavioral and emotional outcomes in children and adolescents after invasive treatment for congenital heart disease. *J Pediatr Surg.* 2008;43:534–539.
7. Palermo TM. Evidence-based interventions in pediatric psychology: progress over the decades. *J Pediatr Psychol.* 2014;39:752–753.
8. Bellinger DC, Wypij D, Rivkin MJ, et al. Adolescents with d-transposition of the great arteries corrected with the arterial switch procedure: neuropsychological assessment and structural brain imaging. *Circulation.* 2011;124(12):1361–1369.
9. McCusker CG. Recovery versus emergence of neurodevelopmental deficits in congenital heart disease from infancy to 7 years. In: *Paper Presented at the 2nd Annual Cardiac Neurodevelopmental Symposium, Cincinnati, USA*; 2013.
10. Watanabe K, Ogino T, Nakano K, et al. The Rey-Osterrieth complex figure as a measure of executive function in childhood. *Brain Dev.* 2005;27(8):564–569.
11. Bellinger DC. Are children with congenital cardiac malformations at increased risk of deficits in social cognition? *Cardiol Young.* 2008;18(1):3–9.
12. von Rhein M, Buchmann A, Hagmann C, et al. Brain volumes predict neurodevelopment in adolescents after surgery for congenital heart disease. *Brain.* 2014;137(Pt 1):268–276.
13. McCusker CG, Doherty NN, Molloy B, et al. Determinants of neuropsychological and behavioural outcomes in early childhood survivors of congenital heart disease. *Arch Dis Child.* 2007;92:137–141.
14. Sarrechia I, De Wolf D, Miatton M, et al. Neurodevelopment and behavior after transcatheter versus surgical closure of secundum type atrial septal defect. *J Pediatr.* 2014;166. [Epub ahead of print].
15. Majnemer A, Limperopoulous C, Shevell M, Rohlicek C, Rosenblatt B, Tchervenkov C. A new look at outcomes of infants with congenital heart disease. *Pediatr Neurol.* 2009;40(3): 197–204.
16. Volpe JJ. Encephalopathy of congenital heart disease- destructive and developmental effects intertwined. *J Pediatr.* 2014;164(5):962–965.

17. McCusker CG, Doherty NN, Molloy B, et al. A controlled trial of early interventions to promote maternal adjustment and development in infants with severe congenital heart disease. *Child Care Health Dev.* 2009;36(1):110–117.

18. McCusker CG, Armstrong MP, Mullen M, Doherty NN, Casey F. A sibling-controlled prospective study of outcomes at home and school in children with severe congenital heart disease. *Cardiol Young.* 2013;23(4):507–516.

19. Doherty N, McCusker CG, Molloy B, et al. Predictors of psychological functioning in mothers and fathers of infants born with severe congenital heart disease. *J Reprod Infant Psychol.* 2009;27(4):390–400.

20. Dulfer K, Duppen N, Van Dijk AP, et al. Parental mental health moderates the efficacy of exercise training on health-related quality of life in adolescents with congenital heart disease. *Pediatr Cardiol.* January 2015;36(1):33–40.

21. McCusker CG, Doherty N, Molloy B, et al. A randomized controlled trial of interventions to promote adjustment in children with congenital heart disease entering school and their families. *J Pediatr Psychol.* 2012;37:1089–1103.

22. Morrison ML, Sands AJ, McCusker CG, et al. Exercise training improves activity in adolescents with congenital heart disease. *Heart.* 2013;99:1122–1128.

23. Law EF, Fisher E, Fales J, Noel M, Eccleston C. Systematic review and meta analysis of parent and family-based interventions for children and adolescents with chronic medical conditions. *J Pediatr Psychol.* 2014;39:866–886.

24. Barkley RA. Distinguishing sluggish cognitive tempo from ADHD in children and adolescents: executive functioning, impairment, and comorbidity. *J Clin Child Adolesc Psychol.* 2013;42(2):161–173.

25. Marino BS, Lipkin PH, Newburger JW, et al. American Heart Association Congenital Heart Defects Committee, Council on Cardiovascular Disease in the Young, Council on Cardiovascular Nursing, and Stroke Council. Neurodevelopmental outcomes in children with congenital heart disease: evaluation and management: a scientific statement from the American Heart Association. *Circulation.* August 28, 2012;126(9):1143–1172.

26. Stehl M, Kazak A, Alderfer M, et al. Conducting a randomised clinical trial of a psychological intervention for parents/caregivers of children with cancer shortly after diagnosis. *J Pediatr Psychol.* 2009;34:803–816.

27. Wade S, Walz N, Carey J, et al. A randomized trial of teen online problem solving efficacy in improving caregiver outcomes after brain injury. *Health Psychol.* 2012;31:767–776.

Index

Printed in the United States
By Bookmasters